W0107663

Jean-Michel Jolion
Walter G. Kropatsch (eds.)

Graph Based Representations
in Pattern Recognition

Computing
Supplement 12

Springer-Verlag Wien GmbH

Univ.-Prof. Dr. Jean-Michel Jolion
Laboratoire Reconnaisance de Formes et Vision,
INSA, Lyon, France

Univ.-Prof. Dr. Walter G. Kropatsch
Pattern Recognition and Image Processing Group,
Institute of Automation, University of Vienna, Austria

© 1998 Springer-Verlag Wien
Originally published by Springer-Verlag/Wien in 1998

Typesetting: Asco Trade Typesetters Ltd., Hong Kong

Graphic design: Ecke Bonk

Printed on acid-free and chlorine-free bleached paper

SPIN: 10674209

With 76 Figures

Library of Congress Cataloging-in-Publication Data

Graph based representations in pattern recognition / Jean-Michel
Jolion, Walter G. Kropatsch, eds,
    p.   cm. — (Computing. Supplement ; 12)
    Based on a workshop held in Lyon, France, April 1997.
    ISBN 978-3-211-83121-2   ISBN 978-3-7091-6487-7 (eBook)
    DOI 10.1007/978-3-7091-6487-7
    1. Optical pattern recognition. 2. Graph theory—Data processing.
I. Jolion, Jean-Michel. II. Kropatsch, W. (Walter) III. Series:
Computing (Springer-Verlag). Supplementum ; 12.
TA1650.G73   1998
006.4'2—dc21   98–23058
                                                CIP

ISSN 0344-8029
ISBN 978-3-211-83121-2

Preface

The rapid increase in computing power allows the use of more complex tools for Pattern Recognition and Image Analysis. Graph theory is such a framework. Along the overall processes from a stimulus to its interpretation, graphs are used for several distinct tasks: hierarchical graphs for image segmentation and for control of perceptual strategies, graph matching for recognition and image understanding, graph manipulation for clustering, conceptual graphs for representation of relational and structural knowledge, involving time explicitly represented in the graphs.

IAPR/TC-15 "Graph based Representations" was created during the meeting of the IAPR governing board in Vienna on 27 August 1996. IAPR/TC-15 held its first workshop on Graph based Representation (GbR'97) on April 17th and 18th, 1997 in Lyon, France, with 30 participants from 6 countries. This book is a collection of 14 papers selected from the presentations, showing a good mixture of both theory and application and of graph theory and pattern recognition. They are grouped into the following subject areas: Hypergraphs; Recognition and Detection; Matching; Segmentation; Implementation Problems; Representation.

This volume starts with a paper introducing the principles of hypergraphs as a general framework for graph representation and processing. Then follows a group of three papers dealing with the application of graphs to the recognition of features and, more particularly, linear features (like edges) in a digital image. As an example of the mixture of both theory and application, the next three papers are concerned with matching graphs. It is demonstrated how matching can be computed even on large graphs. Then we come back to application with three papers on graph based segmentation of images or sequence of images by means of regions. Using a nice framework is of interest, of course, but insufficient if you cannot implement it on a computer. Two papers deal with this issue (which results in the TC-15 work on software, see below). Finally, two papers introduce some new representation ideas for neighbourhood graphs which are very famous in pattern recognition community.

The lively discussions and the many new contacts established as well as the activities planned for the future demonstrate that the subject of both the TC and the first workshop fell on fertile grounds and will continue to grow. As an example, some software packages and bibliography entries are available and can be accessed through http://www.prip.tuwien.ac.at/TC15/. These pages are regularly updated and reflect the activities within TC-15. An electronic mailing list is also available for anybody who wants to contribute to TC-15. (In order to access to this mailing list, send a mail to J.M. Jolion (jolion@rfv.insa-lyon.fr).

Another intensive TC-15 workshop is scheduled for 1999 in Austria. Join us!

January 1998   *Jean-Michel Jolion, Walter G. Kropatsch*

Contents

Hypergraphs

Demko, D.: Generalization of Two Hypergraphs. Algorithm of Calculation of the Greatest Sub-Hypergraph Common to Two Hypergraphs Annotated by Semantic Information 1

Recognition and Detection

Englert, R., Cremers, A. B., Seelmann-Eggebert, J.: Recognition of Polymorphic Patterns in Parameterized Graphs for 3D Building Reconstruction 11

Tupin, F., Mangin, J.-F., Pechersky, E., Nicolas, J.-M., Maître, H.: A Graph-Based Representation to Detect Linear Features 21

Salotti, M.: Edge Detection as Finding the Minimum Cost Path in a Graph ... 33

Matching

Cordella, L. P., Foggia, P., Sansone, C., Vento, M.: Subgraph Transformations for the Inexact Matching of Attributed Relational Graphs 43

Shearer, K., Bunke, H., Venkatesh, S., Kieronska, D.: Efficient Graph Matching for Video Indexing 53

Wendling, L., Desachy, J.: Isomorphism between Strong Fuzzy Relational Graphs Based on k-Formulae 63

Segmentation

Jahn, H.: A Graph Structure for Grey Value and Texture Segmentation 73

Brun, L., Domenger, J.-P., Braquelaire, J.-P.: Discrete Maps: a Framework for Region Segmentation Algorithms 83

Bertolino, P., Ribas, S.: Image Sequence Segmentation by a Single Evolutionary Graph Pyramid 93

Implementation Problems

Kropatsch, W. G., Burge, M., Ben Yacoub, S., Selmaoui, N.: Dual Graph Contraction with LEDA 101

Ducourthial, B., Constantinescu, G., Mérigot, A.: Implementing Image Analysis with a Graph-Based Parallel Computing Model 111

Representation

Pailloncy, J.-G., Jolion, J.-M.: The Frontier-Region Graph 123

Mottl, V. V., Blinov, A. B., Kopylov, A. V., Kostin, A. A.: Optimization Techniques on Pixel Neighborhood Graphs for Image Processing 135

Computing Suppl 12, 1–9 (1998)

© Springer-Verlag 1998

Generalization of Two Hypergraphs.
Algorithm of Calculation of the Greatest Sub-Hypergraph Common to Two Hypergraphs Annotated by Semantic Information

Ch. Demko, La Rochelle

Abstract

The problem of the search of the greatest sub-hypergraph common to two hypergraph is a combinational search problem. The algorithm of exploration of all solutions has an exponential complexity according to the size of the two hypergraphs. In this article, we present methods to minimize the necessary time for the calculation of these solutions. These methods constitute a general framework of the study of these problems and can be used in many areas such natural language, pattern recognition or organic chemistry.

AMS Subject Classification: 68 Computer Science.

Key words: Hypergraphs, relational matching, A* algorithm.

1. Introduction

In the past literature, a lot of authors have used graphs to represent information contained in images and subgraph isomorphism to compare them. We present an extension of these solutions by the use of hypergraphs. Hypergraphs can add a *higher* dimension to graph representation in image processing. It allows to define highly informative relations. For example, we can consider that a ternary relation exists between two regions and their frontier.

So, if we choose to represent information in images by the use of hypergraph techniques, we have to solve the problem of the calculation of the greatest sub-hypergraph common to two hypergraphs. It consists in finding the greatest sub-structures to the two hypergraphs.

The goal of this article is to propose a formal description of

- hypergraph representation
- calculation of the greatest sub-hypergraph common to two hypergraph

Described methods can serve as basis to solve problems evoked by the conceptual graph theory, by structural pattern recognition in artificial vision [6], by

2. Oriented Hypergraph

The theory of oriented hypergraph [3] spreads that oriented graph. In the theory of graph, binary edges can be directed or not. In the theory of the oriented hypergraph, binary edges become n-ary relationships.

Formally, an oriented hypergraph is a couple (V, R) where:

- V is the set of vertices of the hypergraph.
- R is a subset of V^i, $i \in N$

The oriented hypergraph ask more graphic primitive than graph. Vertices are represented by a rectangle and relationships by an ovoid. Each arc of a relationship is annotated by a integer expressing the position of the vertex in the relationship. We introduce a matrix notation to represent an oriented hypergraph. Let (V, R) be an oriented hypergraph. (V, R) can be represented by a matrix of integers $(M_{vr})_{v \in V, r \in R}$ such that M_{vr} (cf. Table 1) represents the signature of the vertex v in the relationship r.

$$M_{vr} = \sum_{i=0}^{a(r)-1} \delta_{v,r_i} 2^i$$

where δ represents the Kronecker symbol, r_i represents the vertex i of the relationship r and $a(r)$ the arity of the relationship r.

For any signature matrix M representing a hypergraph (V, R), M verifies

$$\forall r \in R, \log_2 \left(1 + \sum_{n \in V} M_{nr} \right) = a(r)$$

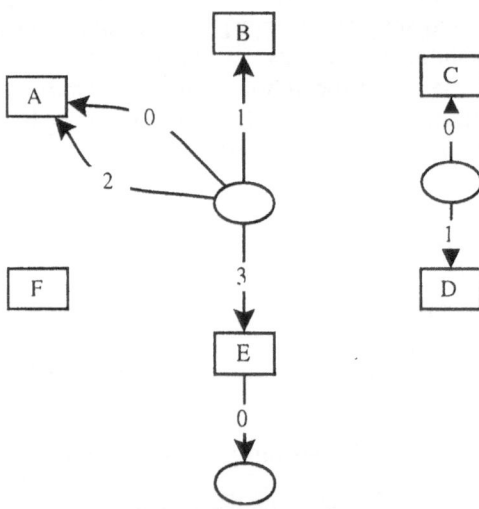

Figure 1. Graphic representation of the hypergraph $(\{A, B, C, D, E, F\}, \{(A, B, A, E), (C, D), (E)\})$

Table 1. Matrix representation of ($\{A, B, C, D, E, F\}$, $\{(A, B, A, E)$, (C, D), $(E)\}$)

	A	B	C	D	E	F
(A,B,A,E)	5	2	0	0	8	0
(C,D)	0	0	1	2	0	0
(E)	0	0	0	0	1	0

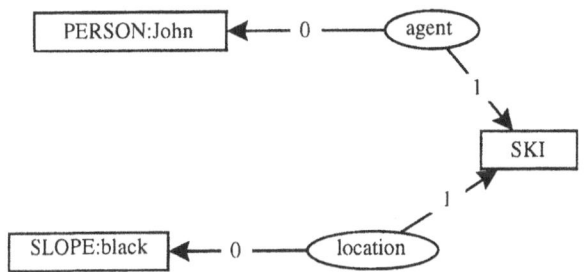

Figure 2. Representation of the sentence *John skis on a black slope*

The addition of semantic information on an oriented hypergraph allows to represent symbolic knowledge. A general formalism has been proposed by J. F. Sowa [5]; it allows to represent symbolic information contained in graph:

- to each vertex v, one associates a function $i(v)$ defining the semantic information supported by the vertex v.
- to each relationship r, one associates a function $i(r)$ defining the semantic information supported by the relationship r.

The conceptual hypergraph of Fig. 2 is defined by ($\{V1, V2, V3\}$, $\{R1 = (V1, V2), R2 = (V3, V2)\}$) and by $i(V1) = $ Person:John, $i(V2) = $ SKI, $i(V3) = $ SLOPE:black, $i(R1) = $ agent and $i(R2) = $ location.

2.1. Notions of Sub-Hypergraph

Notions of sub-hypergraph exposed in this paragraph do not take into account semantic information. We will spread these definitions later. Presented definitions allow to pose the problem of the calculation of the greatest sub-hypergraph common to two hypergraph in a formal manner.

2.1.1. Definition of Sub-Hypergraph

Let a hypergraph (V, R) and a set W such that $W \subset V$. The sub-hypergraph of (V, R) fathered by W is the hypergraph (W, S) where $S = R \cap W^i, i \in N$.

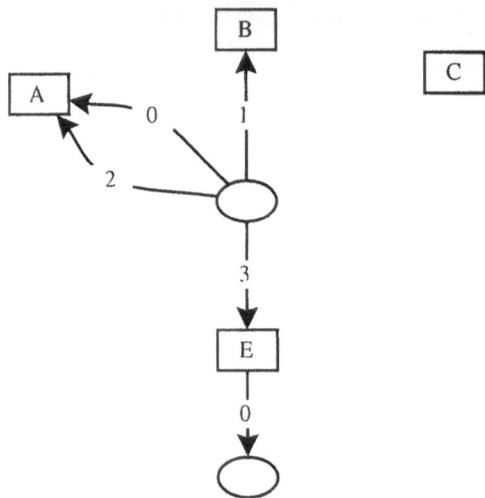

Figure 3. Sub-hypergraph fathered by $\{A, B, C, E\}$ from the hypergraph $(\{A, B, C, D, E, F\}, \{(A, B, A, E), (C, D), (E)\})$

One sub-hypergraph corresponds therefore to the withdrawal of some vertices of the hypergraph.

2.1.2. Definition of Partial-Sub-Hypergraph

Let a hypergraph (V, R) and two set W and S such that $W \subset V$ and $S \subset R$. The partial sub-hypergraph $\sigma((V, R), W, S)$ fathered by W and S is the hypergraph (W, T) such that $T = S \cap W^i, i \in N$.

A partial sub-hypergraph corresponds therefore to the withdrawal of vertices and relationships of the hypergraph.

2.1.3. Definition of the Operation of Generalization Between Two Hypergraphs

The operation of generalization between two hypergraph H_1 and H_2 (noted $\otimes(H_1, H_2)$) calculates the set of the hypergraph H such that H is a partial sub-hypergraph of H_1 and H_2.

2.1.4. Definition of the Maximal Generalization Operation Between Two Hypergraphs

The notion of maximality of the generalization can father several definitions [6]. The chosen one maximize the number of relationships of the partial hypergraph

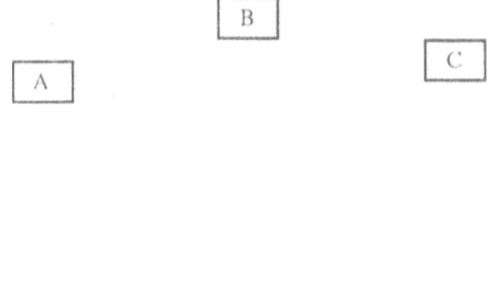

Figure 4. Partial sub-hypergraph of $(\{A, B, C, D, E, F\}, \{(A, B, C, E), (C, D), (E)\})$
fathered by $\{A, B, C, E\}$ and by $\{(C, D), (E)\}$

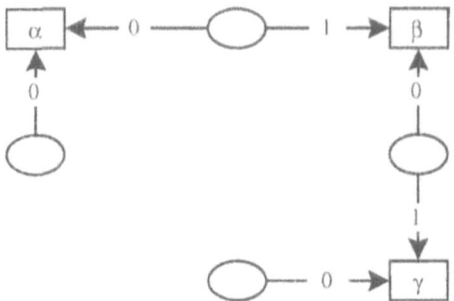

Figure 5. Hypergraph $(\{\alpha, \beta, \gamma\}, \{(\alpha), (\gamma), (\alpha, \beta), (\beta, \gamma)\})$

common to H_1 and H_2. It is to note that the calculation of the maximal generalization between two hypergraph does not give a unique solution.

The maximal generalization $\bar{\otimes}(H_1, H_2)$ between two hypergraph H_1 and H_2 is
the subset of maximal partial sub-hypergraph of H_1 and H_2. It is therefore con-
stituted by the set of the hypergraph H such that H is a partial sub-hypergraph of
H_1 and H_2 and such that there does not exist a partial sub-hypergraph H' of H_1
and H_2 such that H is a partial sub-hypergraph of H'.

For the calculation of $\bar{\otimes}(H_1, H_2)$, one seeks to make correspond structures of the
two hypergraph by maximizing the size of the common structure.

An exhaustive and possible method to calculate $\bar{\otimes}(H_1, H_2)$ would be to calculate
all the partial sub-hypergraph of H_1 and H_2, then to calculate the intersection of

these two set. Due to the fact of the combinational explosion of this method, it can not be applied in the practice. The solution that we have retained consists in construct sequentially the partial sub-hypergraph common to H_1 and H_2. We propose therefore an algorithm of exploration in a tree.

3. Algorithm of Calculation of the Greatest Sub-Hypergraph Common to Two Hypergraphs

Calculating $\bar{\otimes}(H_1, H_2)$ consists in associating between them vertices of the two hypergraph. The algorithm defines therefore a function A between the set of vertices of a hypergraph to the other. The obtained function correspond to a placement in common of one vertex of each hypergraph. Such a function, that we call association function, is a partial function defined on a part of the set of vertices of the first hypergraph. The possible number of these functions is really big. It corresponds to the number of partial inversible functions of a set compound of p elements to a set compound of q elements $\left(\sum_{i=0}^{\min(p,q)} C_p^i A_q^i \right)$. The demonstration is given in [2].

3.1. Association Function

We can represent an association function A by a set of elements of type (v_1, v_2) where v_1 and v_2 are respectively vertices of hypergraph H_1 and H_2. By abuse of language, one will identify the function A to this set.

At the first step s_0 of the algorithm of calculation of $\bar{\otimes}(H_1, H_2)$, $A^{(s_0)}$ is the empty set. To generalize the algorithm and to take into account natural manner the hypergraph annotated by semantic information, we spread the set of vertices of the hypergraph with an absurd vertex that we note λ. Each hypergraph is virtually composed by an infinity of absurd vertices. For a hypergraph (V, R), λ does not appear in any of the elements of R ($\forall r \in R, M_{\lambda r} = 0$).

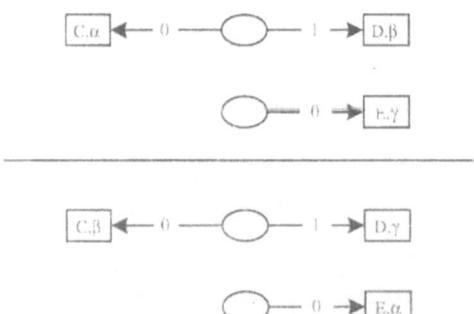

Figure 6. The two maximal generalizations of the hypergraph $(\{A, B, C, D, E, F\}, \{(A, B, C, E), (C, D), (E)\})$ and $(\{\alpha, \beta, \gamma\}, \{(\alpha), (\gamma), (\alpha, \beta), (\beta, \gamma)\})$

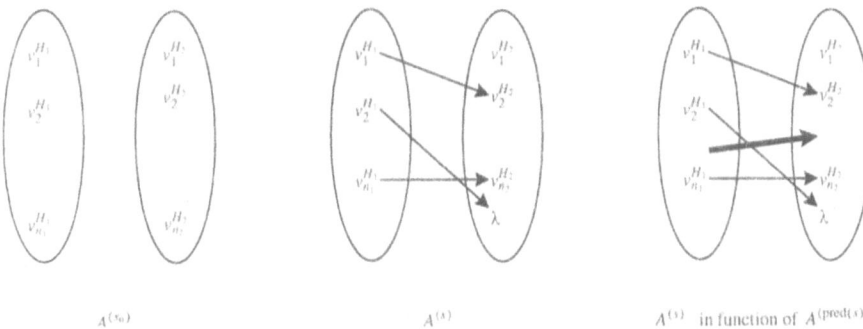

Figure 7. Representation of A where n_1 and n_2 represent respectively the cardinality of V_1 and V_2

At step s, one associates a vertex v_1 of the hypergraph H_1 with a vertex v_2 of the hypergraph H_2 (cf. Fig. 7) where pred(s) is the state predecessor of state s in the exploration tree.

Several solutions are possible to define v_2; each vertex v_1 can be associated with each of vertices v_2 of the hypergraph H_2 that were not been associated to none vertex of the hypergraph H_1 if $i(v_1) \equiv i(v_2)$ and with the absurd vertex λ.

The mechanism of calculation of the best functions of associations has therefore to obey a strategy of search in a tree. Its exploration is realized by a technique of type A∗.

3.2. Cost Function

In our problem, the cost function used at step s is based on the calculation of the set of relationships of H_2 that can be associated to the relationship of H_1.

Let two relationships r_1 and s_1 of the hypergraph H_1. The set of relationship of hypergraph H_2 that can be associated to r_1 and to s_1 are equal $(P(r_1) = P(s_1))$, or disjoint $(P(r_1) \cap P(s_1) = \varnothing)$ [2].

The function P can be seen as a many-to-one of the set of relationships of the hypergraph H_1 to the set of parts of relationships of the hypergraph H_2. $P^{(s_0)}$ is defined by:

$$P^{(s_0)}(r_1) = \{r_2 \in R_2, i(r_1) \equiv i(r_2)\}$$

More, it is possible to define $P^{(s)}(r_1)$ from $P^{(\text{pred}(s))}(r_1)$. Let M_1 and M_2 the signature matrix (cf. Section 2) respectively associated to H_1 and H_2 and (v_1, v_2) the association couple added to $A^{(\text{pred}(s))}$ ($A^{(t)} = A^{(\text{pred}(s))} \cup \{(v_1, v_2)\}$):

$$P^{(s)}(r_1) = \{r_2 \in P^{(\text{pred}(s))}(r_1), M_{1_{v_1 r_1}} = M_{2_{v_2 r_2}}\}$$

From the many-to-one P, it is possible to define a bijection Q that associates with a set of relationships of the hypergraph H_1 a set of relationships of the hyper-

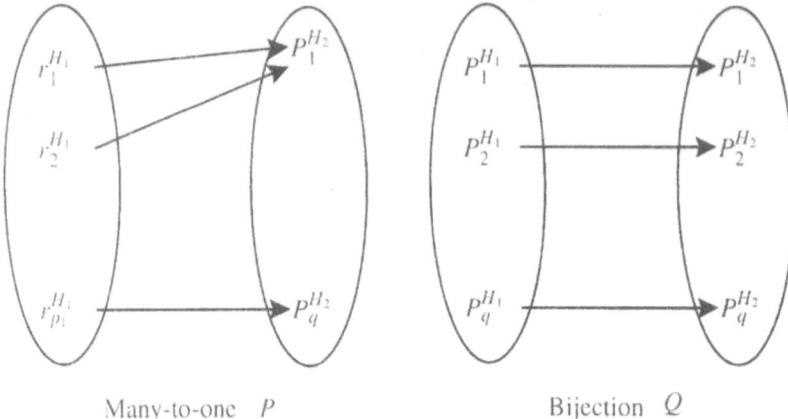

Many-to-one P   Bijection Q

Figure 8. Many-to-one P and bijection Q. p_1 is the cardinality of R_1 and q the number of equivalence classes of R_2 defined by P

graph H_2. Q is therefore a bijection of the set of parts of relationships of the hypergraph H_1 to the set of parts of relationships of the hypergraph H_2 and i.e. a bijection on a set of equivalence classes of R_1 to a set of equivalence classes of R_2.

For a state s of the exploration tree, the maximum number of relationships of the common structure will not be bigger at the end of the exploration tree than $N^{(s)} = \sum_{i=1}^{q^{(s)}} \min(|P_i^{(s)}|, |Q^{(s)}(P_i^{(s)})|)$ [2] where $|.|$ represents the cardinality of a set.

The cost function that we use is

$$c(s) = g(s) + \hat{h}(s)$$

where $g(s)$ is the cost spend from the initial state to the state s and $\hat{h}(s)$ is the estimated cost from the state s to a final state.

$$g(s) = N^{(s_0)} - N^{(s)}$$

$$\hat{h}(s) = (n_1 - p(s))\frac{(N^{(s_0)} - N^{(s)})}{p(s)}$$

$$c(s) = \frac{n_1(N^{(s_0)} - N^{(s)})}{p(s)}$$

where $p(s)$ is the depth of the state s.

Since n_1 is a constant and common to all the states of the exploration tree, $c(s)$ can be simplified to

$$\frac{N^{(s_0)} - N^{(s)}}{p(s)}$$

i.e. a state of the exploration tree is best that an other state of the exploration tree if one estimates that the final number of common relationships will be superior; i.e. the loss of common relationship will be inferior.

More, it is possible, and even desirable, that techniques of exploration in the tree with limited memory or limited time [1] have to be used to avoid the exponential cost of the search; it has, indeed, been shown that this type of problem is NP-complete. Moreover, the use of the estimation of the final number of lost relationship may not conduct to the global optimal solution.

We have seen that the addition of semantic information on relationships allows to constrain the placement in common of relationships of the two hypergraph. The addition of semantic information on vertices allows to restrain the area of the function A for each of these vertices. This functionality serves for example to use a lattice of concepts as in the conceptual graph theory. The addition of semantic information reduces in fact the width of the exploration tree and this to all depths of the tree.

4. Conclusion

We have developed this algorithm in CommonLisp in the framework of a contribution of management of the context for an automatic comprehension system of the language. The formalism of representation that we had chosen was articulated around the conceptual graph theory. The operation of calculation of the greatest sub-hypergraph common to two common hypergraph allowed us to calculate a coefficient of intersection which was used to lift ambiguities present during the analysis of texts. We think nevertheless that it can find its usefulness in applications linked to structural pattern recognition and artificial vision.

References

[1] Chakrabarti, P. P., Ghose, S., Acharya, A., C. de Sarkar, S.: Heuristic search in restricted memory. Art. Intell. *41*, 197–221 (1989).
[2] Demko, Ch.: Contribution à la Gestion du Contexte pour un Système de Compréhension Automatique de la Langue. Thèse de doctorat, Université de Technologie de Compiègne, Compiègne, France, December 1992.
[3] Gondran, M., Minoux, M.: Graphs and algorithms. New York: Wiley, 1990.
[4] Shapiro, L. G., Haralick, R. M.: Relational matching. Appl. Optics *26*, 1845–1851 (1987).
[5] Sowa, J.: Conceptual structures information processing in mind and machines. Reading: Addison-Wesley, 1984.
[6] Vosselman, G.: Relational matching. Lecture Notes in Computer Science, Vol. 628. Berlin, Heidelberg, New York, Tokyo: Springer, 1992.

Ch. Demko
Université de La Rochelle
Laboratoire d'Informatique et d'Imagerie
Industrielles
Avenue Marillac
F-17042 La Rochelle Cedex 1 FRANCE
e-mail: christophe.demko@univ-1r.fr

Computing Suppl 12, 11–20 (1998)

© Springer-Verlag 1998

Recognition of Polymorphic Patterns in Parameterized Graphs for 3D Building Reconstruction

R. Englert, A. B. Cremers, and J. Seelmann-Eggebert, Bonn

Abstract

An approach for the recognition of polymorphic patterns by subgraph isomorphism computation of parameterized graphs will be presented. Parameterized graphs (short: p-graphs) are extensions of undirected graphs by parameter vectors at the nodes and edges. We will define p-graphs and basic concepts of subgraph isomorphism computation for p-graphs. A bottom-up algorithm for p-subgraph isomorphism computation according to a given search graph and a template graph will be described. Since the enumeration of all induced p-subgraphs require tremendous effort, we will propose four pruning mechanisms in order to reduce the size of the search space. The computed subisomorphisms provide useful background knowledge for 3D building reconstruction in order to alleviate the occurring ambiguities, namely occlusions, inverse mapping, and noise. Using an extensive suburban scene it will be shown how polymorphic patterns can be recognized. Finally, we will estimate and discuss an upper bound complexity of the algorithm according to the application.

AMS Subject Classifications: 05C30, 05C70, 68U07, 68T10, 90C35.

Key words: Parameterized graphs, Subgraph matching, Complexity estimation, 3D object reconstruction.

1. Introduction

The recognition and classification of polymorphic patterns for scene analysis is an important endeavor [2]. In this paper an approach for recognizing polymorphic patterns by subgraph isomorphism computation of parameterized graphs will be presented. Although algorithms for isomorphism computation of planar graphs with polynomial cost exist, e.g. [10], subgraph isomorphism computation requires an enumeration and matching of all subgraphs, and is hence exponential in time. We will provide an approach which uses geometrical and topological information at the nodes and edges, so-called *parameter vectors*, and thus can speed-up subgraph isomorphism computation. The information is used in order to guide the search and to prune the search space. An extension of graphs by parameter vectors (short: p-graphs) and concepts of p-subgraph isomorphism will be defined in Section 2. A breadth-first search algorithm for the enumeration of all induced p-subgraphs will be described in Section 3. Since the exhaustive search requires tremendous effort, we will provide several pruning mechanisms in order to reduce computational cost. The approach will be evaluated in the frame of 3D building reconstruction (Section 4) using a large set of 3D building surfaces of the suburban area Oedekoven close to Bonn which is represented as parameterized graphs. Polymorphic patterns of roofs which are significant due to given background knowledge will be computed. Finally, an upper bound complexity estimation of the algorithm presented will be given in Section 5 by taking the application into account.

2. Basic Notions

In this section we will define *parameterized graphs* and *p-subisomorphic* graphs. Basic concepts of graph theory can be found in, e.g. [8]. Central to the presented approach are *parameter vectors* which are based on *intervals*.

Definition 1 (interval). *An* INTERVAL OF THE REAL AXIS *(short:* INTERVAL *)* $[l, r]$ *is a set of points x with* $l \leq x \leq r$, *whereas l is the* LEFT INTERVAL DELIMITER *and r is the* RIGHT INTERVAL DELIMITER. *If* $l = r$ *holds, then the interval consists of exactly one point and is called* POINT INTERVAL. *Furthermore,* $\Box$ *is the* EMPTY INTERVAL, *otherwise it is* NON-EMPTY.

Note, in this paper we will only consider *closed*, non-empty interval. This definition can be easily expanded by *half-open* and by *open* intervals.

Definition 2 (parameter vector). *A* PARAMETER VECTOR $\vec{a} = (a_1, \ldots, a_n)$ *is a list of n non-empty intervals of the real axis, whereas* a_i *is called* COMPONENT OF $\vec{a}$.

Parameter vectors enable one to express geometrical and topological information of objects, e.g. the gradient of a line segment in the three-dimensional Euclidean space. Examples of parameter vectors are $\vec{a} = ([-3.0, 5.0])$ which represents an interval of the length 8.0 units and $\vec{b} = ([5.0, 5.0])$ which denotes a point interval.

Definition 3 (parameterized graph). *A* PARAMETERIZED GRAPH $G_P = (V, \zeta, E, \xi)$ *(short:* P-GRAPH*) is an undirected graph, where each node* $v \in V$ *has a parameter vector* $\zeta(v)$ *and each edge* $e \in E$ *has a parameter vector* $\xi(e)$. *The* NEIGHBORHOOD $\mathcal{N}(V')$ OF A NODE SET $V' \subseteq V$ *is* $\mathcal{N}(V') \stackrel{def}{=} \{v \in V \setminus V' | \exists w \in V' : \{v, w\} \in E\}$ *and the* DEGREE $\delta(V')$ *is* $|\mathcal{N}(V')|$.

Definition 4 (connected). *A p-graph* $G_P = (V, \zeta, E, \xi)$ *is called* CONNECTED *iff* $G = (V, E)$ *is connected.*

P-graphs represent patterns of objects and thus we are primarily interested in connected p-graphs. Since nodes of p-graphs contain parameter vectors, we assume that p-graphs do not entail self-loops. Analogously, p-graphs must not have parallel edges. An example of p-graphs is depicted in Fig. 4. Our goal is to extract patterns of p-graphs and in order to be able to maintain the information we need a mechanism which puts two parameter vectors into a relation to each other.

Definition 5 (IS SUBSUMED BY, $\preceq$). *Given two parameter vectors* $\vec{a} = (a_1, \ldots, a_n)$ *and* $b = (b_1, \ldots, b_n)$. $\vec{a}$ IS SUBSUMED BY $\vec{b}$, *written* $\vec{a} \preceq \vec{b}$, *iff for* $a_i = [a_{il}, a_{ir}]$ *and* $b_i = [b_{il}, b_{ir}]$ *hold* $b_{il} \leq a_{il}$ *and* $a_{ir} \leq b_{ir}$ *for* $i = 1, \ldots, n$.

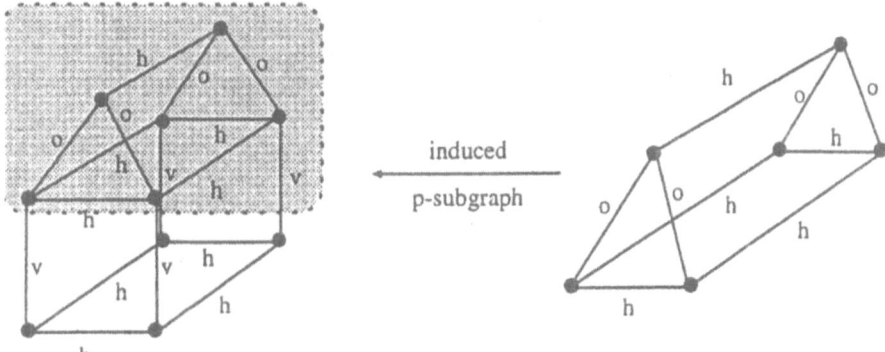

Figure 1. The p-graph (right side) is an induced p-subgraph of the p-graph on the left side (the lower letters h = horizontal, v = vertical, and o = oblique denote the gradient of the line segments)

Examples for the subsumption of parameter vectors are the following: if $\vec{a} = ([-3.0, 5.0])$ and $\vec{b} = ([-4.0, 6.0])$ then $\vec{a} \preceq \vec{b}$ and $\vec{b} \not\preceq \vec{a}$ hold; if $\vec{a} = ([-1.0, 1.0])$ and $\vec{b} = ([-1.0, 1.0])$ then $\vec{a} \preceq \vec{b}$ and $\vec{b} \preceq \vec{a}$ hold, since both parameter vectors are *equivalent*; and if $\vec{a} = ([5.0, 5.0])$ is a point interval and $\vec{b} = ([-4.0, 6.0])$ then $\vec{a} \preceq \vec{b}$ and $\vec{b} \not\preceq \vec{a}$ hold.

Subgraphs of p-graphs have to take relations of parameter vectors of nodes and edges into consideration. The notion of an *induced p-subgraph* enables one to extract a subgraph from a p-graph maintaining its information given by parameter vectors. Note, given a mapping $f : X \to Y$, then the mapping $f|M : M \to Y$, with $M \subset X$, is called the *restriction of f to M*.

Definition 6 (p-subgraph). *A* P-SUBGRAPH *$G'_P = (V', \zeta', E', \xi')$ of a parameterized graph $G_P = (V, \zeta, E, \xi)$ is a connected p-graph with $V' \subseteq V$, $E' \subseteq E \cap \binom{V'}{2}$, $\zeta' = \zeta|V'$, and $\xi' = \xi|E'$. G'_P is called* INDUCED P-SUBGRAPH *OF G_P, iff $E' = \binom{V'}{2} \cap E$.*

As an example for an induced p-subgraph see Fig. 1.

Definition 7 (UNION, $\dot{\cup}$). *Given two induced p-subgraphs $G'_P = (V', \zeta', E', \xi')$ and $G''_P = (V'', \zeta'', E'', \xi'')$ of a p-graph $G_P = (V, \zeta, E, \xi)$, with $V' \cap V'' = \varnothing$. Then G'_P* UNION *G''_P, written $G'_P \dot{\cup} G''_P$, denotes the p-graph $G_{\dot{\cup}} = (V_{\dot{\cup}}, \zeta_{\dot{\cup}}, E_{\dot{\cup}}, \xi_{\dot{\cup}})$, with $V_{\dot{\cup}} = V' \cup V''$, $= \zeta_{\dot{\cup}} = \zeta|V_{\dot{\cup}}$, $E_{\dot{\cup}} = \binom{V' \cup V''}{2} \cap E$, and $\xi_{\dot{\cup}} = \xi|E_{\dot{\cup}}$.*

Definition 8 (P-ISOMORPHIC, $\simeq_P$). *A p-graph $G_P = (V, \zeta, E, \xi)$ is* P-ISOMORPHIC *to a p-graph $G'_P = (V', \zeta', E', \xi')$, written $G_P \simeq_P G'_P$, iff there exist bijections $\phi_V : V \to V'$ and $\phi_E : E \to E'$, with $\phi_E(\{v, w\}) = \{\phi_V(v), \phi_V(w)\}$ for all edges $\{v, w\} \in E$, with $\zeta(v) \succeq \zeta'(\phi_V(v))$ for all nodes $v \in V$, and with $\xi(e) \succeq \xi'(\phi_E(e))$ for all edges $e \in E$.*

As consequence a node v is p-isomorphic to a node v', written $v \simeq_P v'$, if for their embedding p-graphs $G_P = (\{v\}, \zeta|\{v\}, \varnothing, \varepsilon_f)f : \varnothing \to \varnothing$ and $G'_P = (\{v'\}, \zeta'|\{v'\}, \varnothing, \varepsilon_f)$ holds $G_P \simeq_P G'_P$, whereas ε_f denotes the so-called *empty function*. For

```
P₁ = ∅
for all vₜ ∈ Vₜ do
    for all v_S ∈ V_S do
        if [vₜ, v_S, φ_V, φ_E] then
            P₁ = P₁ ∪ {[vₜ, v_S, φ_V, φ_E]}
        end if
    end for
end for
```

Figure 2. Computation of P_1

practical reasons we will abbreviate $G_P \simeq_P G_P'$ by the quadruple $[G_P, G_P', \phi_V, \phi_E]$, and analogously $v \simeq_P v'$ will be abbreviated by the quadruple $[v, v', \phi_V, \phi_E]$. The concepts of an *admissible extension* of induced p-subgraphs and *p-subisomorphic* graphs are in the core of this approach.

Definition 9 (ADMISSIBLE EXTENSION). *Given two induced p-subgraphs $G_P = (V, \zeta, E, \xi)$ and $G_P' = (V', \zeta', E', \xi')$ of G and two nodes $v \notin V$ and $v' \notin V'$. The node pair (v, v') is an* ADMISSIBLE EXTENSION *of the pair (G_P, G_P'), iff $G_P \cup G_v$ is p-isomorphic to $G_P' \cup G_{v'}$, whereas G_v, resp. $G_{v'}$, is the embedded p-graph of v, resp. v'.*

Definition 10 (P-SUBISOMORPHIC, $\sim_P$). *A p-graph $G_P = (V, \zeta, E, \xi)$ is P-SUBISO-MORPHIC to a p-graph $G_P' = (V', \zeta', E', \xi')$, written $G_P \sim_P G_P'$, iff G_P is p-isomorphic to an induced p-subgraph G_P^i of G_P', whereas the upper index i denotes the number of nodes of G_P^i. G_P is called* TEMPLATE GRAPH *and G_P'* SEARCH GRAPH.

As an example for a p-subismorphism see Fig. 1. The p-graph on the right side is p-isomorphic to a p-subgraph of the graph on the left side (indicated by the gray shaded face), since it subsumes all the parameter vectors.

3. P-Subgraph Isomorphism Computation

The algorithm is depicted in Fig. 3 and its initialization in Fig. 2. As input a template graph $T = (V_T, \zeta_T, E_T, \xi_T)$ and a search graph $S = (V_S, \zeta_S, E_S, \xi_S)$ are taken. The main idea is to perform a breadth-first search by traversing the p-graphs along all incident edges given a start node. Breadth-first search graph traversal has been discussed in [9] and is adopted by the use of admissible extensions (cf. Fig. 3). The mechanism of an admissible extension-based graph traversal is described in the following proposition (a weaker version of this proposition has been depicted in [4]).

Proposition 11. *Given induced p-subgraphs $T_1 = (V_{T_1}, \zeta_{T_1}, E_{T_1}, \xi_{T_1})$, $T_2 = (V_{T_2}, \zeta_{T_2}, E_{T_2}, \xi_{T_2})$, $S_1 = (V_{S_1}, \zeta_{S_1}, E_{S_1}, \xi_{S_1})$, $S_2 = (V_{S_2}, \zeta_{S_2}, E_{S_2}, \xi_{S_2})$, of $T = (V_T, \zeta_T, E_T, \xi_T)$, resp. $S = (V_S, \zeta_S, E_S, \xi_S)$, with $[T_1, S_1, \phi_{V_1}, \phi_{E_1}]$ and $[T_2, S_2, \phi_{V_2}, \phi_{E_2}]$, and assume that $V_{T_1} \cap V_{T_2} = \emptyset$, $V_{S_1} \cap V_{S_2} = \emptyset$ and $\forall v_1 \in V_{T_1}, v_2 \in V_{T_2}: \{v_1, v_2\} \in$*

$$
\begin{aligned}
&\text{for } i = 2, \ldots, |V_T| \text{ do} \\
&\quad \text{for all } [G_T = (V_T, \zeta_T, E_T, \xi_T), G_S, \phi_V, \phi_E] \in P_{i-1} \text{ do} \\
&\qquad P_i = \varnothing \\
&\qquad \text{for all } (v_T, v_S) \text{ admissible extension of } (G_T, G_S) \text{ do} \\
&\qquad\quad P_i = P_i \cup \{[G_T \,\dot\cup\, v_T, G_S \,\dot\cup\, v_S, \phi_V', \phi_E']\} \\
&\qquad \text{end for} \\
&\quad \text{end for} \\
&\text{end for} \\
&\text{for } i = 1, \ldots, |V_T| \text{ do Store } P_i \text{ end for}
\end{aligned}
$$

Figure 3. Breadth-first search algorithm for the enumeration of induced p-subgraphs. $A = B$ denotes the assignment of B to A

E_T iff $\{\phi(v_1), \phi(v_2),\} \in E_S$ and $\xi_T(\{v-1, v_2\}) \succeq \xi_S(\{\phi_{V_1}(v_1), \; \phi_{V_2}(v_2)\})$ hold, then $[T_1 \,\dot\cup\, T_2, S_1 \,\dot\cup\, S_2, \phi_{V_\cup}, \phi_{E_\cup}]$, with

$$
\phi_{V_\cup}(v) \stackrel{def}{=}
\begin{cases}
\phi_{V_1}(v), & v \in V_{T_1} \\
\phi_{V_2}(v), & v \in V_{T_2}
\end{cases}
\tag{1}
$$

and

$$
\phi_{E_\cup}(\{v_1, v_2\}) \stackrel{def}{=}
\begin{cases}
\phi_{E_1}(\{v_1, v_2\}), & \{v_1, v_2\} \in E_{T_1} \\
\phi_{E_2}(\{v_1, v_2\}), & \{v_1, v_2\} \in E_{T_2} \\
\{\phi_{V_1}(v_1), \phi_{V_2}(v_2)\}, & v_1 \in V_{T_1} \wedge v_2 \in V_{T_2}
\end{cases}
\tag{2}
$$

Proof: We have to show that the following holds:

1. $\phi_{V_\cup}$ and $\phi_{E_\cup}$ are bijections.
2. The p-isomorphism (Definition 8) holds for $T_1 \,\dot\cup\, T_2$ and $S_1 \,\dot\cup\, S_2$.

First we will consider the mapping of the nodes. Since V_{T_1} and V_{T_2} are disjunctive, resp. V_{S_1} and V_{S_2}, and by using the assumption, that ϕ_{V_1} and ϕ_{V_2} are bijective, the mapping $\phi_{V_\cup}$ is also bijective.

Analogously, the same argument holds for the edges $e \in \{E_{T_1} \cup E_{T_2}\}$. Those edges which have exactly one node in V_{T_1} and one node in V_{T_2}, the precondition of the proposition ensures the bijection.

In order to prove the property of p-isomorphism we have to consider two cases:

1. $e \in \{E_{T_1} \cup E_{T_2}\}$
   Then the p-isomorphism follows from $T_1 \simeq_P S_1$ resp. $T_2 \simeq_P S_2$.
2. $e = \{v, w\} \notin \{E_{T_1} \cup E_{T_2}\}$
   Without loss of generality let $v \in V_{T_1}$. By using the requirements of the proposition it follows that $\{\phi_{V_1}(v), \phi_{V_2}(w)\} \in E_S$ and $\xi(\{v, w\}) \succeq \xi(\{\phi_{V_1}(v), \phi_{V_2}(w)\})$. Hence, the preconditions of Definition 8 hold. ■

However, for large search graphs the search effort can be tremendous. In order to alleviate the search space exploration we will provide so-called pruning mechanisms, which allow to cut off ambiguous solutions as early as they are recognized. For reasons of clearness the pruning mechanisms SHRINKING SET, SYMMETRICAL SOLUTIONS, and LOOK AHEAD which are described in [6] have been left out. Often template graphs have mirror symmetrical properties (cf. Fig. 4), which enable one to apply also the following pruning mechanism P-SYMMETRIC PARTITION: a mirror symmetrical template graph can be partitioned into two equal, induced p-subgraphs. One of the resulting graphs with a reduced set of nodes is subsequently used as template graph. This reduction diminishes the search effort significantly.

Definition 12 (P-SYMMETRIC PARTITION). *Given a template graph* $T = (V_T, \zeta_T, E_T, \xi_T)$ *and a partition* $V_T = V_1 \cup V_2$ *of the node set with* $|V_1| = |V_2|$. $(V_1, V_2, \phi_V, \phi_E)$ *with* $\phi_V : V_1 \rightarrow V_2$ *and* $\phi_E : \binom{V_1}{2} \cap E_T \rightarrow \binom{V_2}{2} \cap E_T$ *is a P-SYMMETRIC PARTITION of* T *if the following conditions hold:*

1. $[G_1, G_2, \phi_V, \phi_E]$ *and* $[G_2, G_1, \phi_V, \phi_E]$, *where* G_1, *resp.* G_2, *denotes the p-subgraph induced by* V_1, *resp.* V_2.
2. $\forall \{v, w\} \in E_T, v \in V_1, w \in V_2 : w = \phi_V(v)$

Suppose a template graph with a p-symmetric partition $(V_1, V_2, \phi_V, \phi_E)$ is given. Then it is only required to process V_1, resp. V_2. Obviously, this improvement can be generalized for partitions consisting of more than two parts, e.g. a circle of length three, which reduces cost significantly.

4. Application: Polymorphic Pattern Computation of 3D Building Surfaces

Our application is based on a three-dimensionally modeled suburban area consisting of 1846 building cluster [5] from which the polyhedral surfaces were computed and represented as p-graphs forming search graphs. The nodes contain as information their degree and the edges contain as information their gradient, their length, and knowledge about the existence of an adjacent, horizontal, orthogonal edge. The parameter vectors represent geometrical and topological information which allow to distinguish the templates and their nodes and edges. Many of the search graphs represent complex building cluster which contain nested or closely interlocking buildings and which comprise detailed information of the roof structure.

The task is to compute relevant building models for photogrammetric building reconstruction [1]. Roofs are most significant in building reconstruction from images [7], and hence relevant building models contain at least a part of a roof. As template graphs served a saddleback roof and a hip-roof (Fig. 4). The algorithm (Fig. 3) took as input the template and search graphs and extracted from each search graph the maximal induced p-subgraph given the template graph. The extracted relevant patterns were grouped together by taking their number of nodes into account. The frequency of occurrence of connected patterns with n nodes of the used templates are depicted in Table 1. The results conclude that the

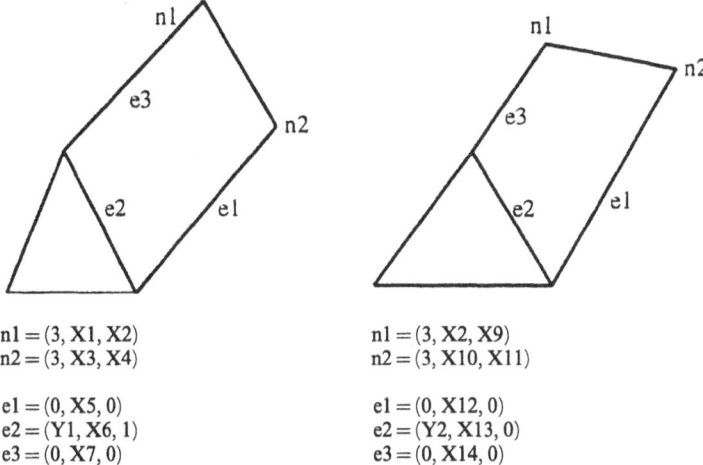

n1 = (3, X1, X2)
n2 = (3, X3, X4)

e1 = (0, X5, 0)
e2 = (Y1, X6, 1)
e3 = (0, X7, 0)

n1 = (3, X2, X9)
n2 = (3, X10, X11)

e1 = (0, X12, 0)
e2 = (Y2, X13, 0)
e3 = (0, X14, 0)

Figure 4. Template graphs with parameter vectors: A saddleback roof and a hip-roof

Table 1. Results of maximal induced p-subgraph computation
using the templates saddleback roof and hip-roof

	Number of nodes					
Template	1	2	3	4	5	6
Saddleback roof	0	24	0	1044	0	893
Hip-roof	0	60	0	7322	28	56

p-subgraphs of the appropriate templates with four nodes occur most frequent. Most extraordinary is the frequency of the recurring results in total 893 times with regard to the computed p-isomorphic graphs of the saddleback roof template. In conclusion for automatic building extraction from images these are the most important structures serving as building models which we will further investigate (cf. Section 6).

5. Upper Bound Complexity Estimation of the Application

In this section an upper bound complexity estimation of the application (cf. Fig. 3) will be given. Given a template graph $T = (V_T, \zeta_T, E_T, \xi_T)$ and a search graph $S = (V_S, \zeta_S, E_S, \xi_S)$, then the sets $P_i = \{[T_i = (V_{T_i}, \zeta_{T_i}, E_{T_i}, \xi_{T_i}),$ $S_i = (V_{S_i}, \zeta_{S_i}, E_{S_i}, \xi_{S_i}), \phi_{V_T}, \phi_{E_T}]\}$, $i = 1, \ldots, |V_T|$, contain induced p-subgraphs T_i (resp. S_i) of T (resp. S) with i nodes. Initially, $P_1 = \{(v_T, v_S)|v_T \in V_T, v_S \in V_S, v_T \simeq_P v_S\}$.

At each step the algorithm examines the neighborhood of the currently considered induced p-subgraph $G = (V', \zeta', E', \xi')$. The degree of V' can be estimated by summing up the degrees of all nodes $v \in V' : \delta(V') \le \sum_{v \in V'} \delta(v)$. Since G is connected, it holds $|E'| \ge |V'| - 1$. Therefore at least $|V'| - 1$ edges are connecting nodes of V'. Each edge increases the degree of two nodes and thus

$$\delta(V') \le \left(\sum_{v \in V'} \delta(v) \right) - 2 \cdot (|V'| - 1) = \sum_{v \in V'} (\delta(v) - 2) + 2. \tag{3}$$

On the other hand, if a graph has $|V'|$ nodes, then maximal $|V \setminus V'|$ nodes can be neighborborrowed to nodes of V', whereas each path contains at least one edge which is incident to a node of $V \setminus V'$:

$$\delta(V') \le |V \setminus V'|. \tag{4}$$

Concluding, we get $\delta(V') \le min\{\sum_{v \in V'}(\delta(v) - 2) + 2, |V \setminus V'|\}$. The application provides template graphs and search graphs with an average node degree of three. Hence, $\delta(V')$ can be simplified to $\delta(V') \le min\{|V'| + 2, |V \setminus V'|\}$. Then the estimated upper bound of $\delta(V')$ improves the upper bound estimation

$$|P_{|V_T|}| \le |P_1| \cdot (|V_T| - 1)! \cdot \frac{(|V_S| - 1)!}{(|V_S| - |V_T| - 1)!} \tag{5}$$

which has been shown in [6] as following, whereas $i = |V_{T_i}| = |V_{S_i}|$:

$$|P_{i+1}| \le \sum_{G \in P_i} min\{i + 2, |V_T| - i\} \cdot min\{i + 2, |V_S| - i\} \tag{6}$$

$$= |P_i| \cdot min\{i + 2, |V_T| - i\} \cdot min\{i + 2, |V_S| - i\} \tag{7}$$

since all summands are independent from G. Consequently, the cardinality of the set $P_{|V_T|}$ uses the upper bound estimation for $|P_i|$ recursively and thus it can be bounded by

$$|P_{|V_T|}| \le |P_1| \cdot \prod_{i=1}^{|V_T|-1} min\{i + 2, |V_T| - i\} \cdot min\{i + 2, |V_S| - i\} \tag{8}$$

Without loss of generality we assume that $|V_T|$ is even and $|V_T| \le |V_S|$. If $i \le \frac{|V_T|}{2}$, then from Eq. 8 follows $min\{i + 2, |V_S| - i\} = i + 2$ by using that $i + 2 \le |V_S| - i$ iff $i \le \frac{|V_S|}{2} - 1$ with regard to $|V_T| \le |V_S|$. This equation is used in order to split the product into two products as following:

$$|P_{|V_T|}| \le |P_1| \cdot \prod_{i=1}^{\frac{|V_T|}{2}-1} (i + 2)^2 \prod_{i=\frac{|V_T|}{2}}^{|V_T|-1} (|V_T| - i) \cdot (|V_S| - i) \tag{9}$$

$$= |P_1| \cdot \frac{1}{4} \cdot \left(\left(\frac{|V_T|}{2} + 1 \right)! \right)^2 \cdot \left(\frac{|V_T|}{2} \right)! \cdot \frac{(|V_S| - \frac{|V_T|}{2})!}{(|V_S| - |V_T|)!} \tag{10}$$

$$= \frac{1}{4} \cdot |P_1| \cdot \left(\left(\frac{|V_T|}{2} + 1 \right)! \right)^3 \cdot \frac{1}{\frac{|V_T|}{2} + 1} \cdot \frac{\left(|V_S| - \frac{|V_T|}{2} \right)!}{(|V_S| - |V_T|)!} \tag{11}$$

$$= \frac{1}{2} \cdot |P_1| \cdot \frac{1}{|V_T| + 2} \cdot \left(\left(\frac{|V_T|}{2} + 1 \right)! \right)^3 \cdot \frac{\left(|V_S| - \frac{|V_T|}{2} \right)!}{(|V_S| - |V_T|)!} \tag{12}$$

Consequently, the upper bound complexity estimation is improved to

$$|P_{|V_T|}| \leq \frac{1}{2} \cdot |P_1| \cdot \frac{1}{|V_T| + 2} \cdot \left(\left(\frac{|V_T|}{2} + 1 \right)! \right)^3 \cdot \frac{\left(|V_S| - \frac{|V_T|}{2} \right)!}{(|V_S| - |V_T|)!}. \tag{13}$$

Let us highlight an example for the improvement: average sizes for template graphs are $|V_T| = 12$ nodes and for search graphs are $|V_S| = 150$ nodes. By using Eq. 5 $B_{old} = 5.45936e + 36$ matching tests have to be performed. The new upper bound complexity (Eq. (13)) decreases the search effort to the execution of $B_{new} = 6.60321e + 25$ tests. This improvement compares to a factor of $\frac{B_{old}}{B_{new}} = 8.26774e + 10$.

Note, the pruning mechanisms have not been incorporated into the estimation. However, their embedding into the application reduces computational cost significantly.

6. Conclusion and Future Work

We have defined a representation for patterns which enable one to encode geometrical and topological information. A graph-based mechanism which is headed by the computation of admissible extensions extracts relevant patterns with regard to user-defined templates. Maximal patterns of buildings which contain parts of roofs have been computed. Finally, an upper bound complexity estimation has been refined taking into account properties of p-graphs of buildings.

A large variety of 3D building models as been extracted. Thus the computed building models require a generalization in order to be more compact. Therefore we investigate the application of Machine Learning approaches as shown in [3]. Furthermore, the computational cost of the admissible extension are huge, and hence raising the central question whether there is a structure of the search space which can be further exploited.

Acknowledgments

The authors like to thank W. Förstner for his discussion. Furthermore they thank J. Zimmermann for comments on an earlier version. This research is supported by BMBF/DLR under Grant 01 M 3018 F/6.

References

[1] Braun, C., Kolbe, T. H., Lang, F., Schickler W., Steinhage, V., Cremers, A. B., Förstner W., Plümer, L.: Models for photogrammetric building reconstruction. Comput. Graphics *19*, 109–118 (1995).

[2] Duda, R. O., Hart, P. E.: Pattern classification and scene analysis. New York: J. Wiley, 1973.

[3] Englert, R.: Inducing integrity constraints from knowledge bases. In: Proceedings of the 19th Annual German Conference on Artificial Intelligence (Wachsmuth, L., Rollinger, C. R., Brawer, W., eds.), pp. 77–88. Lecture Notes in Artificial Intelligence, Vol. 981. Berlin Heidelberg New York Tokyo: Springer, 1995.

[4] Englert, R., Cremers, A. B.: Improving reconstruction of man-made objects from sensor images by machine learning. In: Proceedings of the 11th SPIE's International Symposium on Aero-Sense: Integrating Photogrammetric Techniques with Scene Analysis and Machine Vision III (McKeown, D. H., McGlone, G., Jamet, O., eds.), pp. 263–274, SPIE Proceedings: Birmingham (U.K.), 1997.

[5] Englert, R., Gülch, E.: A one-eye stereo system for the acquisition of complex 3D-bound structures. Geo Inf. Syst., *9*, 16–21 (1996).

[6] Englert, R., Seelmann-Eggebert, J.: P-subgraph isomorphism computation and upper bound complexity estimation. Technical Report IAI-TR-97-2, University of Bonn, Institute of Computer Science III, Bonn, Germany, 1997.

[7] Grün, A., Kübler, O., Agouris, P. (eds.): Automatic extraction of man-made objects from aerial and space images. Basel: Birkhäuser, 1995.

[8] Harary, F.: Graph Theory. Reading: Addison-Wesley, 1972.

[9] Papadimitriou, C., Steiglitz, K.: Combinatorial optimization: algorithms and complexity. Englewood Cliffs: Prentice-Hall, 1982.

[10] Ullmann, J. R.: An algorithm for subgraph isomorphism testing. J. ACM *23*, 31–42 (1976).

R. Englert and A. B. Cremers
Institute of Computer Science III
Rheinische Friedrich-Wilhelms-Universität Bonn
Römerstraße 164, D-53117 Bonn
e-mail: {englert|abc}@cs.bonn.edu

J. Seelmann-Eggebert
Institute of Discrete Mathematics
Rheinische Friedrich-Wilhelms-Universität Bonn
Nassestraße 2, D-53113 Bonn
e-mail: seelmann@or.uni-bonn.de

Computing Suppl 12, 21–31 (1998)

A Graph-Based Representation to Detect Linear Features

F. Tupin, Paris, **J.-F. Mangin,** Orsay, **E. Pechersky,** Moscow,
J. M. Nicolas, Paris, and **H. Maître,** Paris

Abstract

Graph-based representations of the scene are well adapted to introduce high-level knowledge in image segmentation. The problem consists then in searching the graph configuration or labeling minimizing some cost function. In the case of local relationships between the graph nodes, the Markovian framework and simulated annealing algorithms provide some answers to this question.

We are interested in this paper in the automatic or semi-automatic detection of linear structures like roads or hydrological networks in satellite radar images. Using a graph of segments and introducing local contextual properties of these networks, a Markov Random Field is defined to perform the detection. Interaction choice relies on a priori knowledge on the usual aspect of the linear objects to detect. Results are presented for real radar images both for road and hydrological networks.

Key words: Graphs, Markov Random Fields, SAR imagery, road detection.

1. Introduction

Graph-based representation of images is becoming a popular tool since it represents in a compact way the structure of a scene to be analyzed and allows for an easy manipulation of sub-parts or of relationships between parts. Therefore, it has been widely used to control the different levels from segmentation to interpretation.

In this paper, a different aspect of graph-based representation is presented which is more related to the detection stage of picture processing, i.e. to the very low-level of the interpretation task. Within this framework, the problem at hand is expressed as the search for the sub-graph which verifies at best a given objective function composed of two different terms: one reflecting the quality of the detection, the second the global conformity of the graph to a priori knowledge we have on it. Taking into account some specific properties of the graph, it is possible to find an elegant solution to the optimal sub-graph detection via a Markovian modeling which allows for a solution in terms of Markov Random Fields on graphs.

The objective of this paper is twofold: at first, we will express the problem of line detection as a problem of graph optimization; secondly, we adapt the complete formalism of MRF (which has been intensively used up to now for regular graphs as for instance square meshes for picture processing) to graph with non-uniform topology.

We will openly place the problem in a precise application context: the detection of a road network in radar imagery, in order to make it more explicit, but its extension to other line-detection problems (as seen in medical image processing or in robot vision) is straightforward and the formalism is also exportable to many other graphs where Markovian conditions apply.

2. Graph Construction

In the case of the detection of the road network, we start from the extraction of some road-segment candidates (actually the easiest to detect), and then try to connect or delete them in an intelligent way. Indeed, although the detected segments have a high probability to belong to the road network, some of them are false detections due to local linear structures, whereas the others really belong to the roads. Therefore, a closing step is necessary. Many approaches have already been proposed on this subject [3, 4, 7, 8]. But these works often deal more with the interpretation problem of high quality images than with the detection problem starting from poor initialization. In the following, we build a graph representation and define the cost function necessary to introduce the common-sense properties of the network.

At a first stage a detection of candidate segments is made either using conventional picture processing tools, or, as it is made here, with specially developed line detectors adapted to SAR images [10, 11]. Among the segments which are detected by this first step, some belong to the real roads and others are false detections. Besides, numerous parts of the roads are not detected. We make now the strong assumption that the road network can be obtained by connecting the right segments and by rejecting the false detections. Therefore, noting S_d the set of the detected segments, we add to S_d the set S'_d of all the reasonable connections between the segments of S_d. We define S'_d in the following way:

$$S'_d = \{M_i^k M_j^l, i \in S_d, j \in S_d \text{ and } i \mathcal{A} j\}$$

i denoting a segment of S_d, M_i^k with $k \in \{1, 2\}$ its extremities, and $i \mathcal{A} j$ the "possible-connection" relationship between segments i and j of S_d. By "possible-connection" relationship, we mean that the closest end-points of the two segments are close enough and the alignment is acceptable. Using this restriction we do not build the complete graph but restrict it to some possible configurations, which of course drastically reduces the complexity[1] (Fig. 1).

---

[1] In our opinion, this condition is very often fulfilled in image processing problems, since a local situation knowledge is often sufficient.

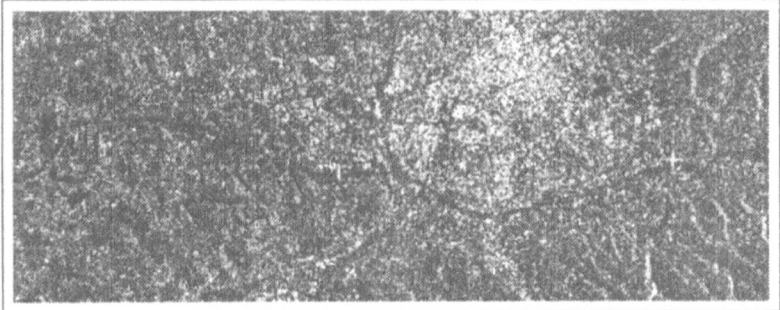

Original SAR image

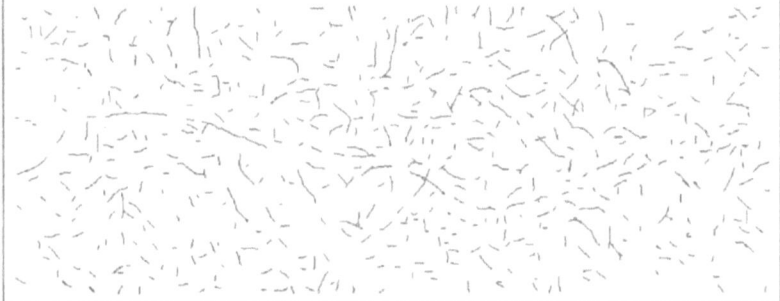

Result of the line detection stage

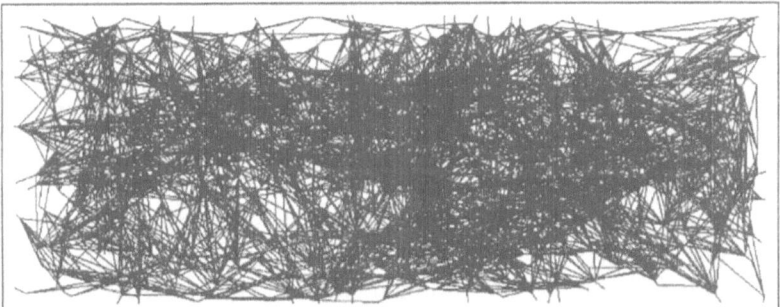

Segments of the graph in the representation image

Figure 1. An example of the graph building on a real SAR image

We define a new set S of segments as $S = S_d \cup S'_d$. S is endowed with a graph structure G: the nodes of the graph are S segments, and an edge links two nodes if the corresponding segments share a same end-point[2] (Fig. 2). With this definition, G is a non planar graph and has a non uniform topology. Some attributes are

---

[2] In fact, each segment in the image representation becomes a node in the graph, and each end-point connecting two segments becomes an edge between the two corresponding nodes.

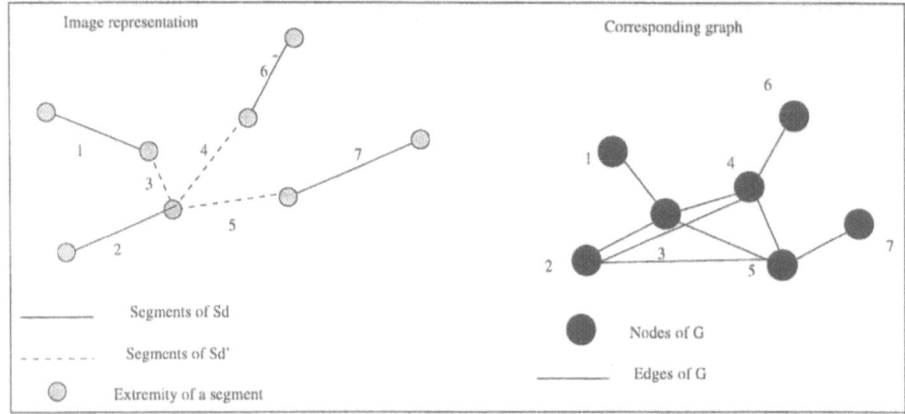

Figure 2. Image and corresponding graph representation

attached to each node and each edge of G resulting in an attributed relational graph. To each node is associated the segment length divided by a maximal length $\mathscr{L}_{max}$ and denoted by $\mathscr{L}_i$ ($\mathscr{L}_i \in [0,1]$), and to each edge between i and j the angle $\mathscr{R}_{ij}$ modulo π between the 2 segments.

3. Problem Definition

The problem of road detection can now be considered as the detection of the "road" and "not-road" nodes in the graph. Therefore, road detection consists in labeling the graph in an appropriate way. In all the following, 1 is the road label and 0 not-road one.

We have now to define an appropriate cost function on this graph to do the road detection. We want to take into account two aspects: first of all, we have to respect the original data and the information they contain; secondly we want to introduce some global knowledge about the common aspect of the road network in satellite images. The Markovian framework is very convenient to introduce both informations.

In this way, a binary random variable L_i is associated to each node $i : L_i = 1$ if i is a "road" segment and $L_i = 0$ otherwise. With N defined as the cardinal of S, $L = (L_1, \ldots, L_N)$ is a random field called label field subsequently. L takes its values in Ω, set of all possible configurations of cardinal 2^N.

Using a Maximum A Posteriori (MAP) estimation, the road detection consists in finding the most probable configuration of L given the observation field D for the segments of S (D is defined below). Therefore we look for the mode of the conditional probability distribution of L given the observations $D = d$, also called posterior distribution $p(L|D)$. Using Bayes rule:

$$p(L|D) = \frac{p(D|L)p(L)}{p(D)}$$

Since $p(D)$ does not depend on the configuration of L, instead of the posterior distribution $p(L|D)$, we can estimate $p(D|L)$ and $p(L)$. The conditional distribution of the observation field given the label field, $p(D|L)$, is estimated using a supervised learning step and permits us to take into account the data. The prior probability distribution of the label field $L, p(L)$, relies on a Markovian model of usual roads and allows us to introduce a priori knowledge.

4. Definition of the Distributions

In this section, based on some assumptions and using the equivalence between Markov Random Field and Gibbs field, we establish that $p(L|D)$ is a Gibbs distribution and we give the form of its associated energy function.

4.1. Conditional Distribution of D Given L

Let us first define the observation process $D = (D_1, \ldots, D_N)$. The responses of the line detector in each pixel along each segment i are computed. The realization of D_i is then given by the mean of these responses[3]. Under the assumption of independence between the D_i and assuming that the conditional distribution of D_i only depends on L_i, we may write:

$$p(D|L) = \prod_{i=1}^{N} p(D_i|L) = \prod_{i=1}^{N} p(D_i = d_i|L_i = l_i) \propto \exp\left(-\sum_{i=1}^{N} V(d_i|l_i)\right)$$

$V(d_i|l_i)$ being called the potential of segment i. The conditional distributions $p(D_i|L_i)$ for $L_i = 0$ and $L_i = 1$ can be learnt from an experiment after a manual segmentation of the roads in some images. The potentials can be deduced from this learning step. Using linear potentials with large areas of transition provides a robust estimation (only one radar image has been used to do the learning step for all the images of the same sensor).

An example on the SAR networks is given Fig. 3. The observation along a segment is defined as the average value of the line detector response [11] at each pixel belonging to the segment. As shown on the Fig. 3a and b, the only discriminating segments are the "not-road" ones, since the human observer is able to connect road segments even in places where the road is not visible using contextual knowledge. Some simple linear potentials (Fig. 3c) have given satisfying results.

---

[3] But another definition could be used depending on the low level detector behaviour (for instance the percentage of pixels whose response is above a fixed threshold).

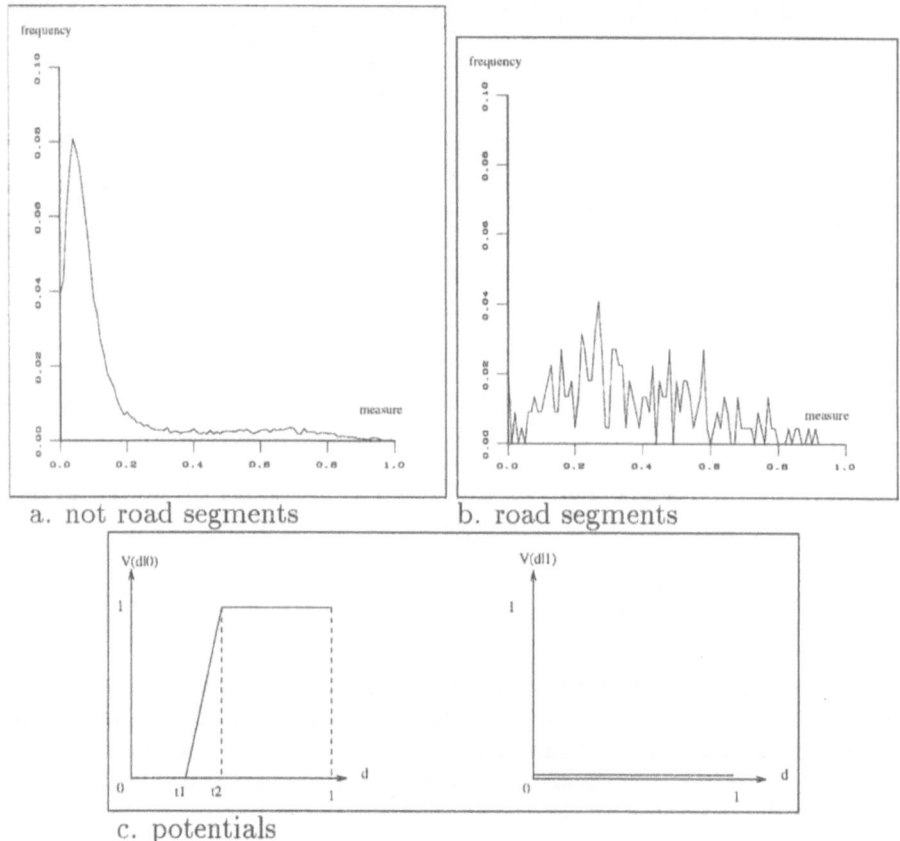

Figure 3. Frequencies obtained using a manual segmentation of the network (**a, b**) and the deduced potentials (**c**)

4.2. Prior Probability Distribution of L

Our knowledge of the usual shape of a road permits us to detect visually the network knowing only a small part of the image (small on the scale of the segments, but large compared to the scale of the pixels). Therefore, we can model L as a Markov Random Field following the constraint:

$$\forall l \in \Omega, p(L = l) > 0 \quad \text{and} \quad p(L_i = l_i | L_j = l_j, j \neq i) = p(L_i = l_i | L_j = l_j, j \in \mathcal{V}_i)$$

$\mathcal{V}_i$ being the neighborhood of the node i (defined in § 3). Using the MRF-Gibbs field equivalence (Hammersley-Clifford theorem [1]):

$$p(L = l) = \frac{1}{Z'} \exp(-U(l))$$

where Z' is a normalizing constant, C denotes the clique set, and $U(l) = \sum_{c \in \mathscr{C}} V_c(l)$. Clique potentials are chosen to express the following a priori knowledge about roads [5]: (i) roads are long (they should almost never stop), (ii) roads have a low curvature, (iii) intersections are rare (i.e. a segment is more often connected to a unique other segment in one of its extremities than to many segments, at least in non urban areas). As a consequence, a road is modeled as an infinite succession of segments with low curvature. The third condition does not forbid crossroads but gives them a lower probability than the connection between only two segments. The flexibility of the Gibbs field framework allows us to construct simple potentials endowing the random field with a probability distribution stemming from this a priori knowledge.

All clique potentials $V_c(l)$ are null except for the cliques of highest order corresponding to the sets of segments sharing the same common extremity for all segments, which turns out to be sufficient for modeling all the interactions between road segments defined above. It is an appropriate choice since we are not interested in the relationship between two segments sharing an extremity, but in the relationship between all the segments sharing the same extremity. The maximal order clique potentials are defined as follows:

$$\forall i \in c, l_i = 0 \Rightarrow V_c(l) = 0 \tag{1}$$

$$\exists! i \in c / l_i = 1 \Rightarrow V_c(l) = K_e - K_{\mathscr{L}} \mathscr{L}_i \tag{2}$$

$$\exists! (i,j) \in c^2 / l_i = l_j = 1, \mathscr{R}_{ij} > \frac{\pi}{2}$$

$$\Rightarrow V_c(l) = -K_{\mathscr{L}} \times (\mathscr{L}_i + \mathscr{L}_j) + K_c \sin \mathscr{R}_{ij} \tag{3}$$

$$\text{in all other cases } V_c(l) = K_i \sum_{i/i \in c} l_i \tag{4}$$

All the parameters which are introduced in the potential definition are connected in a simple way with the three previously expressed road characteristics. First of all, K_e is chosen positive since we want to penalize road extremities (condition (i) Eq. (1)). $K_{\mathscr{L}}$ is also positive since we want to favorize long chains of aligned segments (condition (i) and (ii) Eq. (1) and (2), Fig. 4). As for the third condition, since K_i is positive, many "road" segments connected at a same extremity or not aligned segments is not a favorable configuration (Eq. (4)).

To detect other networks and assuming the fundamental hypothesis expressed at the beginning of Section 2 is verified (i.e. the network can be obtained by connecting the detected segments in an appropriate way), other a priori knowledge can be introduced by using other clique potentials.

From a more usual structural point of view, the a priori model embedded in this Gibbs distribution could be related to a syntactic point of view. Indeed, our approach is similar to relaxation approaches involving local syntactical constraints [2]. Another parallel could be done with stochastic grammars.

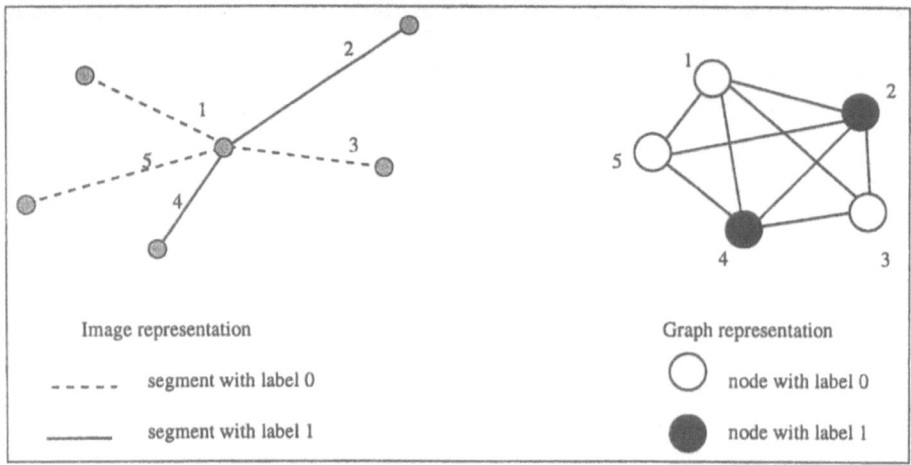

Figure 4. Example of a good road configuration (2 segments with label 1 connected at the same end-point): left in the image representation, right in the graph representation (corresponding to the maximal clique potential defined in Eq. (3))

4.3. Posterior Distribution

Since $p(D|L)$ and $P(L)$ correspond to Gibbs distributions defined on the same graph, so does the global field probability distribution, with the associated energy $U(l|d)$:

$$U(l|d) = \sum_{i=1}^{N} V(d_i|l_i) + \sum_{c \in C} V_c(l)$$

Therefore, the MAP configuration corresponds to the energy minimum. Since the energy function is non-convex, a stochastic minimization algorithm (simulated annealing) has been chosen.

A specific exploration topology is used to overcome convergence problems. Instead of using a very low temperature decreasing rate, sets of segment are considered during the simulated annealing. We have empirically found that considering sequentially sets of three adjacent sites (segments here) gave better and faster results than taking only one site sequentially as it is usually made. This is a way to adapt the exploration topology to the energetic landscape we deal with.

4.4. Parameter Estimation

For many Markovian methods, the problem of parameter estimation is a crucial one. In our case, all the parameters introduced in the potential definition can be set by analyzing some particular configurations.

For instance, we can study when two segments with road labels should be connected. Since we do not want to systematically connect two segments, we deduce after calculations and making some extreme assumptions (for instance about the observation measure or the alignement of the segments), that $K_e + K_{\mathscr{L}}$ should be less than a given threshold depending on the distribution $p(D|L)$.

In the same way, we can analyse when a road configuration for a segment chain has a lower energy than the non-road configuration. The calculation of the energy variation between the two configurations gives us a relationship between $(K_e, K_{\mathscr{L}})$ and the length of the chain, considering once again some extreme assumptions.

5. Results and Conclusion

The proposed method is illustrated on real radar images of ERS-1 both for road and hydrological networks, showing the potential of the method and the difficulties still to solve (for instance, the connection is sometimes incomplete and some roads are detected twice). Both hydrological and road networks are very important features for many applications: automatic registration with other sensor images or maps, cartographic application in very cloudy regions, geomorphologic studies. Besides, the strong speckle noise of the radar images forbid the use of some classical methods. Examples of results are given in the case of the hydrological network in Fig. 5 and of the road or water-channel network (Fig. 6). Even if the results are not perfect, they present a real progress compared to other works on the subject of road or linear feature detection on radar images, since most of them only detect small disconnected road segments [6, 9, 12].

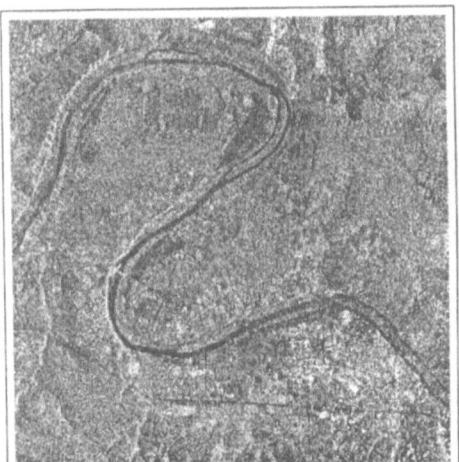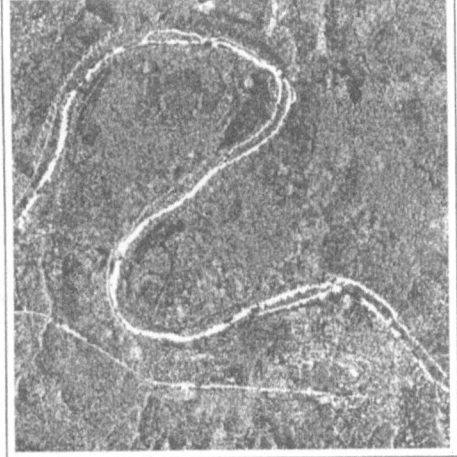

Figure 5. Original image (Mantes-La-Jolie, France) © ESA (left) and the Seine detection result superimposed on the original image (right)

F. Tupin et al.

Figure 6. Original image (Flevoland in The Netherlands) © ESA (above) and the road detection result superimposed on the original image (below)

Besides, using the parameter influence analysis, the intervals in which the parameters have to be taken can be deduced. In this way, the method is almost automatic. Furthermore, the proposed framework is very general and can also be used for other linear features for which some a priori assumptions can be made. Other applications like the detection of objects with a particular shape could also be performed. The fundamental assumption to be fulfilled is that you are able to model the objects to detect using some local properties.

References

[1] Besag, J.: Spatial interaction and the statistical analysis of lattice systems. J. R. Statist. Soc. Ser. B *36*, 192–326 (1974).
[2] Bunke, H.: Attributed programmed graph grammars and their application to schematic diagram interpretation. IEEE Trans. Pattern Analy. Mach. Intell. *4*, 574–582 (1982).
[3] Cox, I. J., Rehg, J. M., Hingorani. S.: A Bayesian multiple-hypothesis approach to edge grouping and contour segmentation. Int. J. Comput. Vision *11*, 5–24 (1993).
[4] David, C., Zucker, S. W.: Potentials, valleys and dynamic global coverings. Int. J. Comput. Vision *5*, 219–238 (1990).
[5] Guérin, P.: Apport des cartes topographiques pour l'analyse de scènes en imagerie aérienne: application à la détection du réseau routier. PhD thesis, Université Paris VII, 1996.
[6] Hellwich, O., Mayer, H., Winkler, G.: Detection of lines in synthetic aperture radar (SAR) scenes. ISPRS Int. Arch. Photogramm. Remote Sensing *31*, 312–320 (1996).
[7] Kass, M., Witkin, A., Terzopoulos, D.: Snakes: Active contours models. Int. J. Comput. Vision *1*, 321–331 (1988).
[8] Lowe, D. G.: Organization of smooth image curves at multiple scales. In: Second International Conference on Computer Vision Florida, USA, pp. 558–567, 1989.
[9] Samadani, R., Vesecky, J. F.: Finding curvilinear features in speckled images. IEEE Trans. Geosci. Remote Sensing *28*, 669–673 (1990).
[10] Touzi, R., Lopes, A., Bousquet, P.: A statistical and geometrical edge detector for SAR images. IEEE Trans. Geosci. Remote Sensing *26*, 764–773 (1988).
[11] Tupin, F., Gouinaud, C., Maître, H., Crettez, J-P., Nicolas, J.-M.: Détection de structures linéaires sur des images ROS. Traitement du Signal *13*, 635–650 (1997).
[12] Wood, J. W.: Line finding algorithms for SAR. R. Signals Radar Establ. (Memorandum 3 841), 1985.

F. Tupin
J.-M. Nicolas
H. Maître
École Nationale Supérieure des
Télécommunications
Département Images
46 rue Barrault
75013 Paris
France
e-mail: tupin@ima.enst.fr maitre@ima.enst.fr

J. F. Mangin
Service Hospitalier Frédéric Joliot
CEA, Orsay
France

E. Pechersky
Institute for Problems of Information
Transmissions 19 Ermolovoj GSP-4
Moscow 101447
Russia

Computing Suppl 12, 33–41 (1998)

Edge Detection as Finding the Minimum Cost Path in a Graph

M. Salotti, Corte

Abstract

We present new ideas to perform contour following using heuristic search strategies. On the one hand, we show that it is interesting to develop the entire graph of the search without defining the goal node: simple graph tools can be used to find closed boundaries or to determine junctions. On the other hand, we show that a cost function with exponential curvature can favor adaptive search strategies, preserving the minimum cost path condition.

Key words: Edge detection, heuristic search.

1. Introduction

Heuristic search strategies have been developed for many applications of artificial intelligence [9]. Martelli applied the A Algorithm to boundary detection in 1972 [6]. Other similar approaches have been proposed, most of them for the detection of boundaries in biomedical applications [5], [12] (an excellent overview is proposed by Ballard and Brown [2]). The main problem lies in the adaptability of the search. In particular, it is difficult to avoid the exploration of small undesirable paths.

We present in the next part the basis idea of heuristic search strategies. Then, we discuss the problem of the representation and we show that it is interesting to develop the entire graph of the search. We also propose a cost function that enables depth first strategies. Finally, some results are presented to illustrate good and bad points of the method.

2. Heuristic Search Strategies

2.1. The Basic Algorithm

In 1972, Martelli showed that the problem of boundary detection can be cast to the problem of finding the minimal cost path in a weighted and directed graph, with positive costs [6].

34   M. Salotti

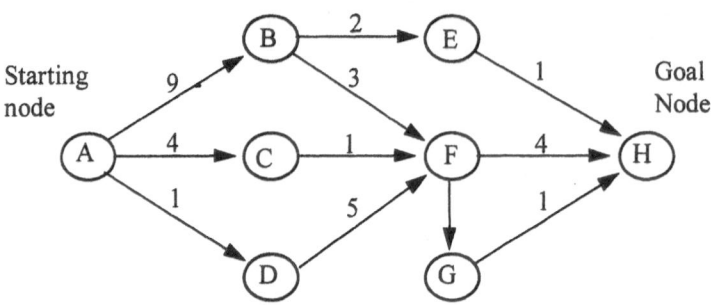

Figure 1. Example of weighted graph

Table 1. Evolution of the search applied to the graph presented in Fig. 1

Step	1	2	3	4	5	6
Open nodes	(A 0)	(B 9) (C 4) (D 1)	(B 9) (C 4) (F 6)	(B 9) (F 5)!!	(B 9) (G 7) (H 9)	(B 9) (H 8)!!
Choice	A	D	C	F	G	H
Best path	A	AD	AC	ACF	ACFG	ADFGH

The basic search is given by Nilsson's algorithm [8]:

1. Expand the start node: put the successors on a list called OPEN with pointers back to the start node.
2. Remove the node Pi of minimum path cost from OPEN. If Pi is a goal node, then stop. Trace back through pointers to find the optimal path. If OPEN is empty, then fail.
3. Else expand node Pi, adding successors on OPEN with the associated path cost, or eventually updating them, with pointers back to Pi. Go to step 2.

An example is presented in Fig. 1 and Table 1. It is interesting to notice that node E is not met during the search, because the cost from A to B is prohibitive. This property is important because it allows us to build dynamically the graph and to take only the best nodes into account.

The A Algorithm is an improvement of the search when the distance to the goal node can be estimated. If g is the cost function of the path from the starting node to a node N, and h is an estimation of the remaining cost between N and the goal node, then $f = g + h$ is a global cost function that can be used instead of g. The solution is guaranteed to be the minimum cost path, provided that h is inferior or at least equal to the exact distance remaining to reach the goal node.

2.2. Representations

Different representations can be chosen: for instance, Martelli proposed to map the nodes to the frontier between two pixels [7]. For convenient reasons, in particular the fact that the result can be displayed as an edge map, we prefer to match directly the nodes with the pixels of the image:

- The starting node can be selected among pixels with high gradient values.
- The successors of a node can be determined by systematically selecting the only three pixel-nodes that allow a curvature of the contour smaller than 45°.
- A critical point is the definition of the ending node. If closed boundaries have to be found, then it is possible to set the starting node as ending node. In this case, if contextual information exists about the shape of objects, then h can be used to constrain the search using the A Algorithm. When finding closed boundaries is not imposed, it could be an interesting idea to develop all the paths until the open list falls to empty. If simple restrictions are made to avoid visiting all pixels (for instance, a node is created only if the local gradient value is superior to a threshold Gmin), the result is a graph of connected edge pixels. Such a graph is a powerful representation of edges: for instance, graph tools can be used to find closed boundaries (special arcs must be added when mark pixels are encountered) or to determine junctions. Moreover, the entire image can be scanned and several graphs can be built from different seeds, pixels-nodes being marked after each search. In this case, an edge map is available and there is a graph for each connected set of edge pixels. In terms of computing time, it is not so bad, providing that an image of the pointers to the nodes is dynamically built to speed up the search in the open list.

2.3. The Local Cost Function

Let $C(n1, n2)$ be the cost of the arc between two nodes $n1, n2$, and let $h(n)$ be the estimated cost to reach the goal node from a node n.

Then, the global cost $f(n)$ of the path going through a node n is given by (1):

$$f(n) = \left(\sum\nolimits_{\text{path}} C(n_i, n_{i+1}) \right) + h(n) \tag{1}$$

Martelli suggested to set $h(n)$ to 0 and to use as local cost an expression roughly equivalent to the following [7]:

$$C(n1, n2) = M - \text{gradient}(n2), \quad \text{where } M = \max_x \{\text{gradient}(x)\} \tag{2}$$

The problem of the heuristic search applied to edge detection is that the graph is dynamically built and it may become very large if using heuristics like expression (2): long paths are quickly expensive and small undesirable paths are explored. Several ideas have been proposed to solve the problem. The selection of the next node in the OPEN list can be made with a depth-first strategy or using a rating function [3], [10], [12]. Lester et al. suggest to take the maximum cost arc of the path instead of the sum. The advantage is that the cost of the path does not grow continuously with depth, so that good paths can be followed for a long time [5]. Since the cost of the path necessary increases with depth if the costs are positive, Ashkar and Modestino proposed a cost function that takes negative values if the arc has a good evaluation [1]. However, if some interesting ideas have been proposed, the property of finding the path with minimum cost is not always guaranteed.

3. Propositions for a Suitable Cost Function

3.1. Exponential Curvature

What should be the shape of the local cost function C?

Let us assume for the moment that the gradient value is the only parameter of the function and that no information is available on the objects to be found. The main idea of the search is that a depth first strategy should be used as long as the gradient value is high. In this case, the values of the cost function should be small enough to make the cost of the entire path smaller than a single cost corresponding to a lower gradient value.

Mathematically, it can be expressed as follows:

Let S be the current path $\{n_1, n_2 \ldots n_k\}$ to reach n_k from the starting node, $k - 1$

$$\forall n \notin S, \forall i \in \{1k\}, \quad \text{Grad}(n) < \text{Grad}(n_i) \rightarrow \left(\sum_{j=1}^{k-1} C(n_j, n_{j+1}) \right) < C(n_i, n) \quad (3)$$

Since only positive costs are to be used, this expression is always true if and only if C is set to 0 for gradient values superior to $\text{grad}(n)$ and to a constant different from 0 otherwise. However, it is important to notice that there is a sum of only k terms. k is simply the depth of the path and in fact the length of the contour. From a practical standpoint, the length of any contour rarely stands 100 or 200 pixels. In this case, it is possible to choose a function that quickly decreases, so that condition (3) is roughly respected. We propose to use a function with exponential curvature, as illustrated in Fig. 2. Let $n1$ and $n2$ be two connected pixels, let x be the gradient value on pixel $n2$ and Gmin a threshold to reject small gradient values, then:

If $x \leq \text{Gmin}$, $\quad C(n1, n2) = \infty \quad$ (or the arc is simply not created).

If $x > \text{Gmin}$, $\quad C(n1, n2) = -1 + e^{c/(x-\text{Gmin})^2} \quad$ ($c > 0$ is used for regularization)

For instance, with Gmin set to 10, and c set to 1, a path of 40 pixels with all gradient values equal to 200 has a smaller cost than a path of 2 pixels with gradient values equal to 50. The search is in fact almost depth first. What is the role of constant c? Its role is to determine the exact shape of the cost function, so that a depth first strategy is more or less favored. When c tends to 0, the cost function

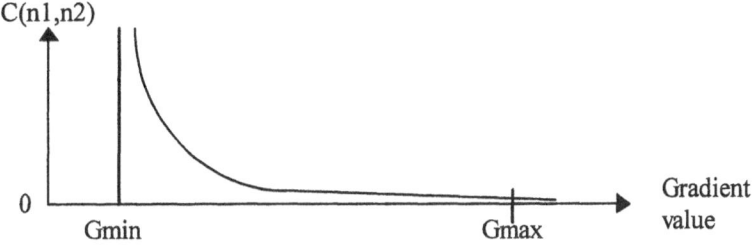

Figure 2. A suitable cost function

tends to a step function and condition (3) tends to be respected with higher probability. Moreover, when c tends to 0, because of the curvature, the maximum arc cost of a path can be taken as a rough approximation of the cost of the entire path. In fact, this approach is similar to the one of Lester et al., except that ours guarantees the minimum cost of the path [5]. In a previous work, we proposed a slightly different function with Gaussian curvature [11]. The function presented here is more appropriate because it is expected to tend to infinity when the gradient tends to Gmin.

3.2. Generalization

Since an exponential curvature is suitable to favor a depth first strategy, a generalization can be made for an arbitrary number of features.

$$C(n1, n2) = \sum_k \alpha_k e^{1/x_k^2} \tag{4}$$

For k features, x_k is an expression of feature k, and α_k determines the associated weight, such that $\sum_k \alpha_k = 1$. Examples of interesting features can be found in previous works [2], [11].

4. Results and Conclusion

Before applying our method, a gradient map is obtained with Deriche operator. Several results are presented, using the input image displayed in Fig. 3:

In Fig. 4, boundaries of black cells have been detected using specific heuristics such as the convexity of the contour or the proximity of dark regions. See [11] for more details, in particular for the definition of the local cost function.

A second group of results is presented in Figs. 5, 6, and 7: we propose to end the search when the Open list falls to empty and to use the function C previously described. Figure 5 shows the pixel-nodes of a single graph (the seed was a pixel with high gradient value in the middle of the image). Figure 6 shows the pixel-nodes after removing branches that are not part of cycles. Edges are rather thin

M. Salotti

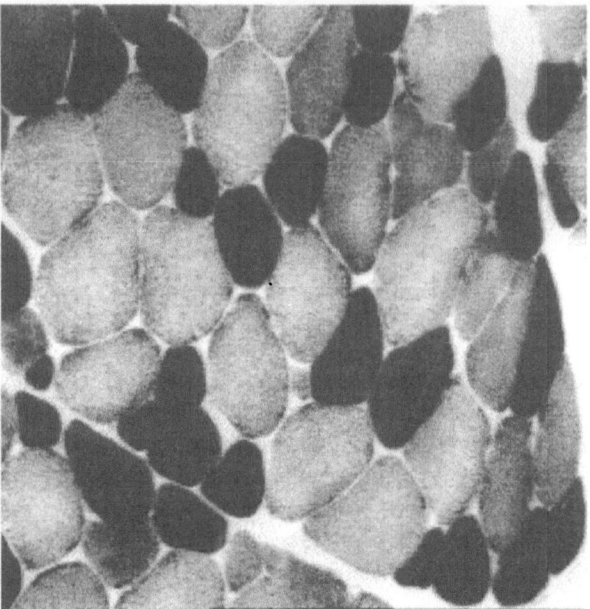

Figure 3. Input image

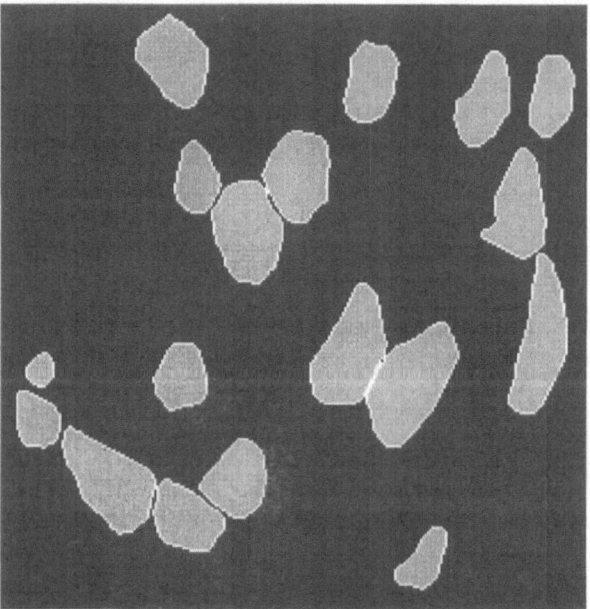

Figure 4. Black cells boundaries obtained using application dependent heuristics

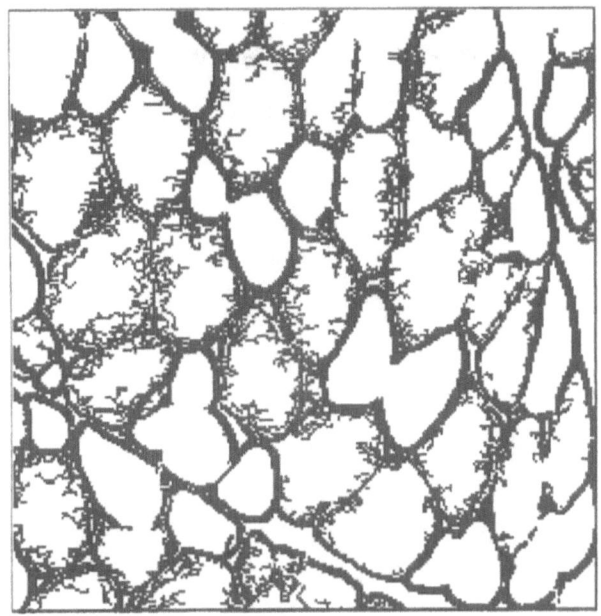

Figure 5. Pixels explored during the search from a single starting node

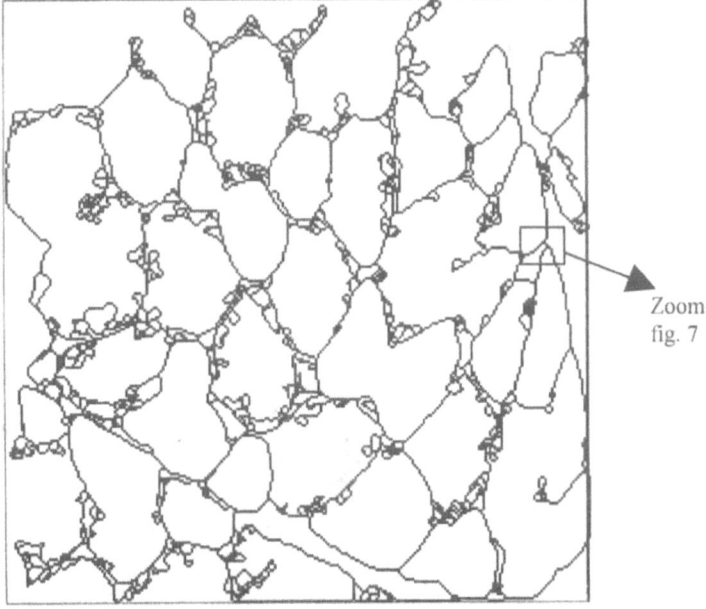

Zoom
fig. 7

Figure 6. Pixels of the graph after removing isolated branches

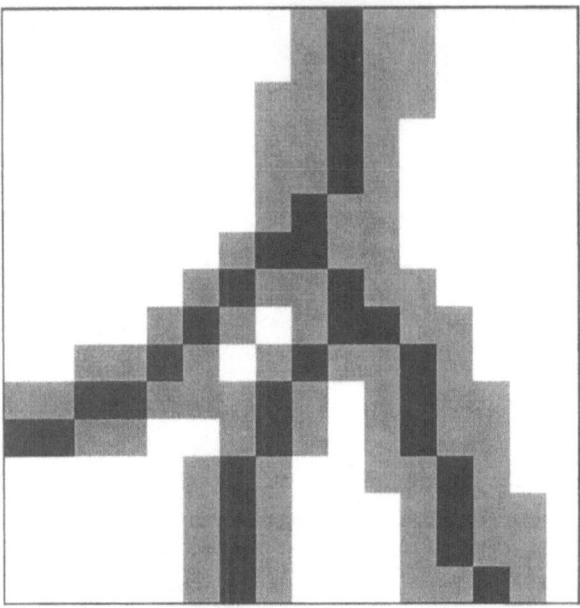

Figure 7. Zoom of Fig. 6. Grey pixels are pixels explored during the search, Black pixels denote pixels selected for the best path from the starting node

because each node has only one predecessor (the one that makes the minimum cost path) and no cycle can be found with pixels located in the border of the edge. Figure 7 is a zoom of the previous one: the pixels of the graph are marked but only the darkest ones belong to cycles. It is interesting to notice the correct localization of junctions.

Our conclusion is that graphs and heuristic search strategies provide rich and powerful tools for edge detection and representation. We are intended to carry on working on this technique and to present more interesting results in a near future.

References

[1] Ashkar, G. P., Modestino, J. W.: The contour extraction problem with biomedical applications. CGIP 7, 331–355, 1978.
[2] Ballard, D. H., Brown, C.: Computer-vision. Englewood Cliffs: Prentice Hall Inc., New Jersey, 1982.
[3] Ballard, D. H., Slansky, J.: A ladder-structured decision tree for recognizing tumors in chest radiographs. IEEE Trans. Comp. 25, 503–513 (1976).
[4] Deriche, R.: Using Canny's criteria to derive a recursively implemented optimal edge detector. Int. J. Comput. Vision 1, 167–187 (1987).
[5] Lester, J. M., Williams, H. A., Weintraub, B. A., Brenner, J. F. Two graph searching techniques for boundary finding in white blood cell images. Comp. Biol. Med. 8, 293–308 (1978).
[6] Martelli, A.: Edge detection using heuristic search methods. CGIP 1, 169–182 (1972).
[7] Martelli, A.: An application of heuristic search methods to edge and contour detection. Comm. ACM, 19, 73–83 (1976).
[8] Nilsson, N. J. S.: Principles of artificial intelligence. Palo Alto: Tioga, 1980.

[9] Pearl, J.: Heuristics: intelligent search strategies for computer problem solving. Reading: Addison-Wesley, 1984.

[10] Persoon, E.: A new edge detection algorithm and its applications in picture processing. CGIP *5*, 425–446 (1976).

[11] Salotti, M., Hatimi, M.: A new heuristic search for boundary detection. In: Lecture Notes in Computer Science *974* (Braccini, C., de Floriani, L., Vernazza, G., eds.) pp. 381–386. Berlin Heidelberg New York Tokyo: Springer, 1995.

[12] Weschler, H., Sklansky, J.: Finding the rib cage in chest radiographs. Pattern Rec. *9*, 21–30 (1977).

Faculté des Sciences et Techniques
BP 52, 20250 Corte
France
e-mail: salotti@univ-corse.fr

Computing Suppl 12, 43 – 52 (1998)

Subgraph Transformations for the Inexact Matching of Attributed Relational Graphs

L. P. Cordella, P. Foggia, C. Sansone, and M. Vento, Naples

Abstract

An inexact matching algorithm for Attributed Relational Graphs is presented: according to it, two graphs are considered similar if, by using a defined set of syntactic and semantic transformations, they can be made isomorphic to each other. The matching process is carried out by using a State Space Representation: a state represents a partial solution of the matching between the graphs, and a transition between two states corresponds to the addition of a new pair of matched nodes. A set of feasibility rules are introduced for pruning states associated to partial matching solutions which do not satisfy the required graphs morphism. Results outlining the computational cost reduction achieved by the method are given with reference to a set of randomly generated graphs.

AMS Subject Classification: 68T10 (68R10).

Key words: Attributed relational graphs, graph isomorphism, inexact matching.

1. Introduction

Structural descriptions of visual patterns can be conveniently put in form of Attributed Relational Graphs (ARGs) [1]. An ARG provides both syntactic information, held by the layout of unlabeled nodes and branches respectively identifying the structural primitive components of the pattern and their relations, and semantic information consisting of the attributes associated to nodes and branches. In real applications, pattern variability is such that samples are seldom identical to prototypes. The variations may be reflected by both syntactic and semantic deformations of the representative graphs, so that pattern recognition can only be achieved by inexact graph matching methods able to manage both these possibilities. The quite few approaches to inexact graph matching proposed in the literature, try to extend the applicability of exact matching methods, by introducing criteria allowing matching in presence of syntactic and/or semantic deformations. In [2], a pattern deformational model is proposed. In its preliminary form only a class of semantic (structure-preserving) deformations is considered and the matching algorithm is based on a graph isomorphism for the syntactic part of the ARG. A generalization of the method, including the possibility of deleting nodes and branches, is proposed in [3]. The algorithm, though powerful enough for some practical applications, is not effective when large variations among the members of a same class may exist. In these cases inexact matching approaches based on the definition of a distance measure between graphs, seem more appropriate. In [4] the distance measure is based on the evaluation of the minimum number of transformations to be applied to one of the compared graphs in order to obtain the other one. Specifically, the distance measure is defined as the recognition cost of ARG nodes plus the cost of a given set of transformations including attribute values modification so as node and branch insertion and deletion. In [1], several algorithms for ARG distance evaluation, in the case that the two graphs are syntactically monomorphic, are described and compared. These algorithms use a search tree to represent the matching process and a set of heuristic criteria for pruning unprofitable branches of the search tree. In a subsequent paper [5] the method is generalized by computing the distance even in case the syntactic parts of the graphs do not satisfy a monomorphism, but only a biunivocal correspondence between the nodes of the graphs holds. Some possible transformations on the branches of the graph are defined, but constraints on their applicability are not considered.

Another relevant problem when matching graphs is that of limiting the computational cost of the process. A classical graph-subgraph isomorphism algorithm which addresses the problem of reducing the matching time is proposed in [6], while the approach presented in [7] is aimed to reduce the overall computational cost when matching a sample graph against a large set of prototypes.

In this paper we define and evaluate an ARG inexact matching algorithm which, using a set of feasibility rules [8], allows on one side to reduce the computational cost of the matching process and on the other side to take into account deformations on syntactic and semantic parts. Peculiar to the method is the fact that the transformations which can be taken into account when matching a sample with a given prototype are only those included in a set of transformations considered applicable to that prototype and singled out during a preliminary training phase. In particular a sample graph is considered similar to one of the prototypes if it is possible to find a set of syntactic and semantic transformations such that the transformed graph sample is isomorphic to the graph of the prototype.

Section 2 describes the matching algorithm and the feasibility rules used to prune the search space in the case of exact graph-subgraph isomorphism. Section 3 introduces the model of the allowed transformations and the extensions of the algorithm to the case of inexact graph-subgraph isomorphism. A discussion of the performance of the graph matching method is finally presented in Section 4 with reference to a set of randomly generated graphs.

2. Exact Graph-Subgraph Matching Algorithm

A matching process between two graphs $G_1 = (N_1, B_1)$ and $G_2 = (N_2, B_2)$ consists in the determination of a mapping M which associates nodes of the graph G_1 to nodes of G_2 and vice versa. As it is well known, different properties can be

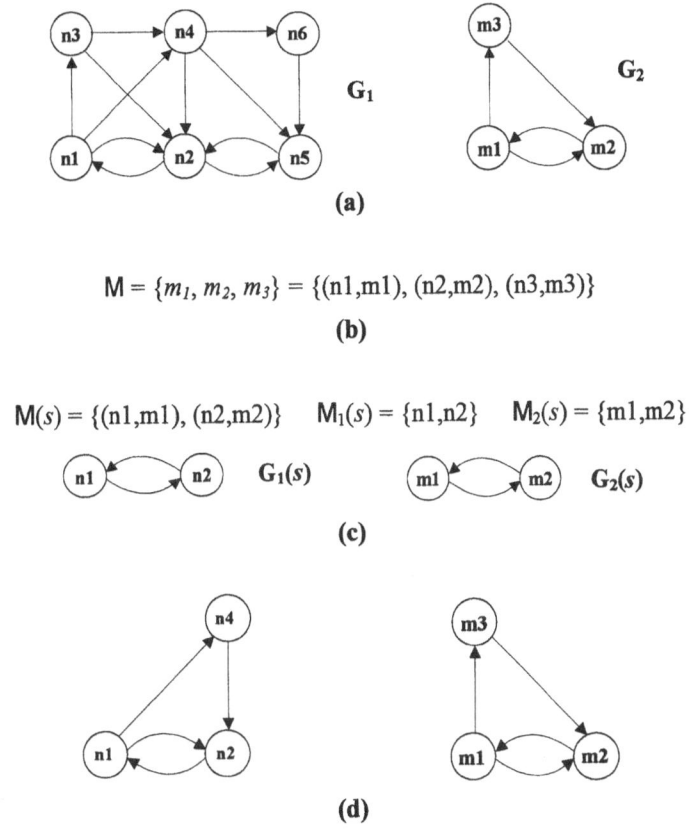

Figure 1. a Two graphs G_1 and G_2, **b** a graph-subgraph solution, **c** a partial mapping solution $M(s)$ and the corresponding subgraphs $G_1(s)$ and $G_2(s)$, **d** the addition of a pair to the state s (transition operator)

required on M and consequently different mapping types can be obtained: monomorphism, strict isomorphism and graph-subgraph isomorphism are the most classical ones.

Generally, the mapping is expressed as the set of ordered pairs (n, m) (with $n \in G_1$ and $m \in G_2$) each representing the mapping of a node n of G_1 with a node m of G_2. Each pair is here denoted as component m_i of the mapping M:

$$M = \{(n, m) \in N_1 \times N_2 \mid n \text{ is mapped onto } m\} = \{m_1, m_2, \dots, m_r\} \qquad (1)$$

The State Space Representation (from now on SSR) can be effectively used to describe a graph matching process, if each state s of the matching process represents a partial mapping solution. A partial mapping solution $M(s)$ is a subset of M, i.e. contains only some components $\{m_1, m_2, \dots, m_k\}$ of M. In Fig. 1 two graphs, a graph-subgraph mapping M and a partial mapping solution are presented. In the following we will denote as $M_1(s)$ and $M_2(s)$ the projection of $M(s)$

onto N_1 and N_2 respectively. In the adopted SSR representation a transition between two states corresponds to the addition of a new pair of matched nodes (see Fig. 1d).

In principle, the solutions to the matching problem could be obtained by computing all the possible partial solutions and selecting the ones satisfying the wanted mapping type (Brute Force approach). In order to reduce the number of paths to be explored during the search, for each state on the path from s_0 to a goal state, we impose that the corresponding partial solution verify some coherence conditions, depending on the desired mapping type. For example, to have a graph-subgraph isomorphism it is necessary that the partial mappings are isomorphisms between the corresponding subgraphs. If the addition of a node pair to the partial mapping produces a state that does not meet the coherence condition, further exploration of that path can be avoided, since it certainly cannot lead to a goal state.

The rationale of our algorithm is that of introducing, given a state s, criteria for foreseeing if s has no coherent successors after a certain number of steps. It is clear that these criteria (feasibility rules) should allow to detect as soon as possible conditions leading to incoherence; in particular, we say that a rule implements a k-look-ahead if, given a state s and a pair (n, m) to be included in s (so obtaining a state s') is allows to establish if all the states reachable from s' in k steps are incoherent. Therefore states which do not satisfy a feasibility rule can be discarded from further expansions.

In Fig. 2 the proposed algorithm is outlined. At each iteration of the outer loop, the algorithm considers the set $P(s)$ of node pairs that can be added to the state s, discarding those pairs which does not satisfy the feasibility rules.

```
INPUT: The graphs G₁ and G₂    OUTPUT: A mapping M
BEGIN
  M(s₀):=∅;  S:={s₀};    (* Initialization *)
  REPEAT
    FOREACH state s contained in S
      Remove s from S
      Determine all the pairs candidate to be included
      in M(s). All the pairs give rise to the set P(s)
      FOREACH candidate pair p∈P(s)
        Evaluate the feasibility rules
        IF the feasibility rules succeed for the inclusion of p
           in M(s)
        THEN add to S the state obtained including p into M(s)
        (* The pairs of P(s) which satisfy the feasibility
         * rules give rise to the set Q(s). Obviously Q(s)⊆P(s)
         *)
      ENDFOR
    ENDFOR
  UNTIL all the nodes of G₂ have been considered OR S is empty
END.
```

Figure 2. The matching algorithm

0-LOOK-AHEAD

R_pred$(s, n, m) \Leftrightarrow$

$\quad (\forall n' \in M_1(s) \cap \mathrm{Pred}_1(n)\, \exists m' \in \mathrm{Pred}_2(m) \mid (n', m') \in M(s)) \wedge$

$\quad \wedge (\forall m' \in M_2(s) \cap \mathrm{Pred}_2(m)\, \exists n' \in \mathrm{Pred}_1(n) \mid (n', m') \in M(s))$

i.e. iff for each predecessor n' of n in the partial mapping, the corresponding node m' is a predecessor of m, and vice versa.

R_succ$(s, n, m) \Leftrightarrow$

$\quad (\forall n' \in M_1(s) \cap \mathrm{Succ}_1(n)\, \exists m' \in \mathrm{Succ}_2(m) \mid (n', m') \in M(s)) \wedge$

$\quad \wedge (\forall m' \in M_2(s) \cap \mathrm{Succ}_2(m)\, \exists n' \in \mathrm{Succ}_1(n) \mid (n', m') \in M(s))$

i.e. iff for each successor n' of n in the partial mapping, the corresponding node m' is a successor of m, and vice versa.

1-LOOK-AHEAD

R_termin$(s, n, m) \Leftrightarrow$

$\quad (\mathrm{Card}(\mathrm{Succ}_1(n) \cap T_1^{in}(s)) \geq \mathrm{Card}(\mathrm{Succ}_2(m) \cap T_2^{in}(s))) \wedge$

$\quad \wedge (\mathrm{Card}(\mathrm{Pred}_1(n) \cap T_1^{in}(s)) \geq \mathrm{Card}(\mathrm{Pred}_2(m) \cap T_2^{in}(s)))$

i.e. iff the number of predecessors (successors) of n that are in $T_1^{in}(s)$ is greater or equal to the number of predecessors (successors) of m that are in $T_2^{in}(s)$

R_termout$(s, n, m) \Leftrightarrow$

$\quad (\mathrm{Card}(\mathrm{Pred}_1(n) \cap T_1^{out}(s)) \geq \mathrm{Card}(\mathrm{Pred}_2(m) \cap T_2^{out}(s))) \wedge$

$\quad \wedge (\mathrm{Card}(\mathrm{Succ}_1(n) \cap T_1^{out}(s)) \geq \mathrm{Card}(\mathrm{Succ}_2(m) \cap T_2^{out}(s)))$

i.e. iff the number of predecessors (successors) of n that are in $T_1^{out}(s)$ is greater or equal to the number of predecessors (successors) of m that are in $T_2^{out}(s)$

2-LOOK-AHEAD

R_new$(s, n, m) \Leftrightarrow$

$\quad (\mathrm{Card}(\mathrm{Pred}_1(n) \cap (N_1 - M_1(s) - T_1(s))) \geq (\mathrm{Card}(\mathrm{Pred}_2(m) \cap (N_2 - M_2(s) - T_2(s)))) \wedge$

$\quad \wedge (\mathrm{Card}(\mathrm{Succ}_1(n) \cap (N_1 - M_1(s) - T_1(s))) \geq (\mathrm{Card}(\mathrm{Succ}_2(m) \cap (N_2 - M_2(s) - T_2(s)))).$

i.e. iff the number of predecessors (successors) of n that are neither in $M_1(s)$ nor in $T_1(s)$ (new nodes) is greater or equal to the number of predecessors (successors) of m that are neither in $M_2(s)$ nor in $T_2(s)$

Figure 3. The feasibility rules for exact graph-subgraph isomorphism

There are two kinds of feasibility rules, respectively regarding the syntax and the semantics of the graphs. The syntactic feasibility rules defined for an exact iso-morphism has been described in [8]; the extension to the case of graph-subgraph isomorphism is shown in Fig. 3.

It is worth noting that the 0-look-ahead rules (R_pred and R_succ) ensure neces-sary and sufficient conditions for the coherence of the successive partial solutions, while the other provide necessary (but not sufficient) conditions and are mainly used for pruning the search graph.

Note that the rules are evaluated with reference to some sets which depend on the considered state s. In particular we have denoted as $T^{out}{}_1(s)(T^{in}{}_1(s))$ and

State s:
$M(s) = \{(n1,m1), (n2,m2)\}$
$M_1(s) = \{n1,n2\}$ $M_2(s) = \{m1,m2\}$
$T^{in}_1(s) = \{n3,n4,n5\}$ $T^{in}_2(s) = \{m3\}$
$T^{out}_1(s) = \{n3,n4,n5\}$ $T^{out}_2(s) = \{m3\}$
$P(s) = P^{out}(s) = \{(n3,m3),(n4,m3),(n5,m3)\}$
$Q(s) = \{(n3,m3),(n4,m3)\}$

(a)

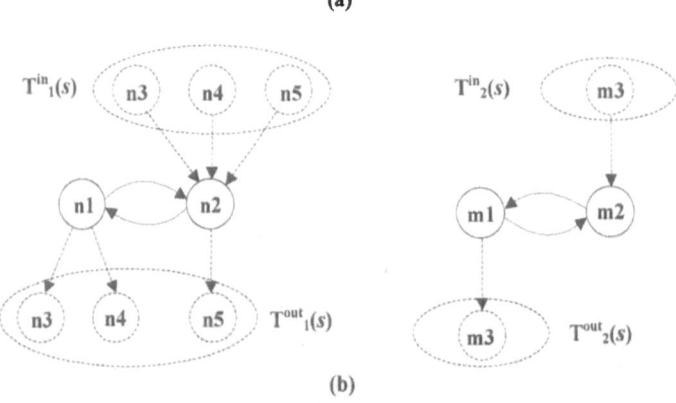

(b)

Figure 4. a A state s, **b** the sets $T_1(s)$ and $T_2(s)$

$T^{out}_2(s)(T^{in}_2(s))$ the sets of outgoing (ingoing) nodes from the two subgraphs $G_1(s)$ and $G_2(s)$, and $T_1(s) = T^{out}_1(s) \cup T^{in}_1(s)$, $T_2(s) = T^{out}_2(s) \cup T^{in}_2(s)$. Moreover, $P(s)$ is the set of pairs (n,m), with n in $T^{out}_1(s)$ and m in $T^{out}_2(s)$, if there are branches outgoing from $G_1(s)$ and $G_2(s)$, otherwise $P(s)$ is the set of pairs (n,m) with n in $T^{in}_1(s)$ and m in $T^{in}_2(s)$. In Fig. 4, a state s and the corresponding sets are given.

Semantic compatibilities can be introduced very easily in the matching process: each time a node of the sample is compared to a node of the prototype to determine if a new pair can be added to the current partial solution, the attributes of the two nodes and of the branches linking them to the nodes already in s are tested for semantic compatibility. Obviously semantic compatibility has to be defined with reference to the specific application domain.

3. Extension of the Method to the Inexact Matching

The exact matching algorithm can be extended by considering both transformations on the structure of the graph and on nodes and branches attributes.

The considered syntactic transformations are the split of a node into a subgraph, the merge of a subgraph into a node and the insertion or deletion of a branch. A split transformation is specified by the 4-tuple (N, B, N', B'), where the set N is made of the only node n to be split, B is the set of branches connected to n, N'

is the set of nodes obtained by splitting n, and B' the set of branches after the transformation. A merge transformation can be likewise specified by a 4-tuple (N, B, N', B') where N now contains more than one node and N' contains the node produced by the transformation.

Transformations implying branch insertion and deletion are specified by the attributes of the branch and of the two nodes connected by it.

Syntactic transformations are taken into account, during the expansion process of the search graph, by generating new states. An example of the SSR expansion in the case of inexact matching is given in Fig. 5. When examining a state in the search process, the algorithm checks if there is a syntactic transformation that can be applied to it. For each applicable transformation, a new state s is added to the SSR graph. In Fig. 5 the states generated by syntactic transformations are not explicitly shown, but their presence is indicated by labeling the state to which the transformation has been applied. The nodes involved in a syntactic transformation are marked, in order to avoid reconsidering them in succesive transformations. In this way we avoid the possible generation of infinite length paths in the search graph and prevent the possibility that a prototype is matched with a too different sample, as a consequence of the repeated application of some transformation.

The conditions a state s has to meet in order that a transformation can be applied, depend on the transformation type. For each type of transformation, only the nodes in $P(s)$ are considered as candidates so as to ensure that, in the next step of the algorithm, the nodes generated by the transformation (or at least some of them) will be tested using the feasibility rules. In this way, if the new paths are fruitless, they will be pruned as soon as possible.

Given a node n_1 of the prototype G_1, not generated by a previous transformation, the condition for applying to n_1 the split transformation (N, B, N', B') is:

$$(n_1 \approx n) \wedge (\forall (n, n') \in B \exists (n_1, n_1') \in B_1 | (n_1, n_1') \approx (n, n')) \wedge$$
$$\wedge (\forall (n', n) \in B \exists (n_1', n_1) \in B_1 | (n_1', n_1) \approx (n', n)) \wedge$$
$$\wedge (\forall (n_1, n_1') \in B_1 \exists (n, n') \in B | (n_1, n_1') \approx (n, n')) \wedge \qquad (2)$$
$$\wedge (\forall (n_1', n_1) \in B_1 \exists (n', n) \in B | (n_1', n_1) \approx (n', n))$$

where the symbol $\approx$ denotes the semantic attribute similarity. If this condition is verified, a new graph is built by deleting the node n_1 together with all the branches connected to it and by inserting the new nodes N' and the branches B'. To properly define the branches that connect N' with the remaining nodes of G_1 an appropriate relabeling is necessary.

In the general case the applicability of a merge transformation is more difficult to detect: the problem is equivalent to that of finding a graph-subgraph isomorphism, although it can be expected that a small size graph is involved. In case of a strict isomorphism, a merge transformation on the prototype can be considered as equivalent to a split transformation on the sample, and its applicability can be

L. P. Cordella et al.

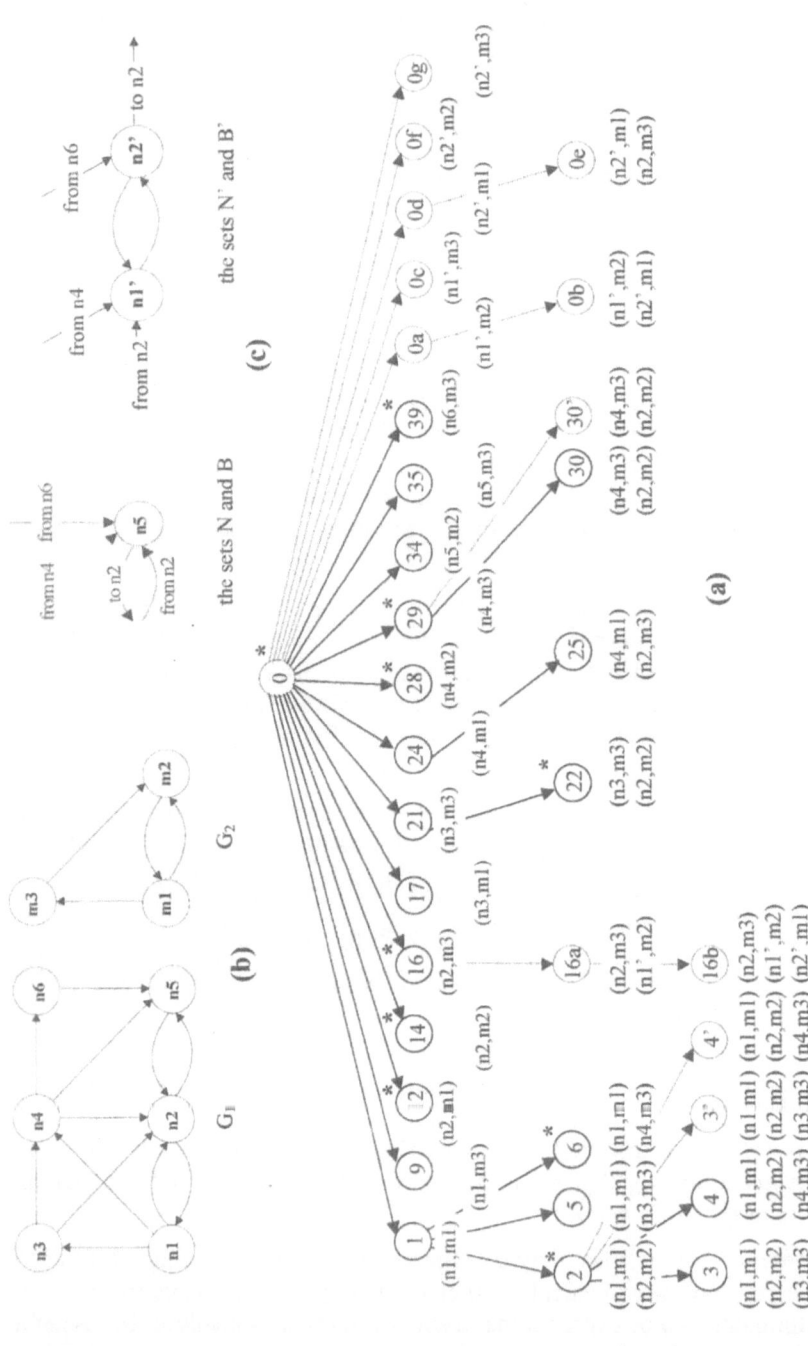

Figure 5. The SSR search graph (**a**) of the inexact graph-subgraph isomorphism between the two graphs G_1 and G_2 (**b**). Dotted lines represent search paths involving the use of the split transformation shown in **c**. The numbering of the nodes is not consecutive because the nodes pruned by the feasibility rules are not shown

tested by using the condition (2), after replacing G_1 with G_2 and exchanging N with N' and B with B'.

Finally, in case of branch insertion and deletion, if (n, n') is the branch to be inserted (deleted), and n_1 and n_1' are two nodes of G_1, the condition for insertion is $(n_1 \approx n) \wedge (n_1' \approx n')$, while the one for deletion is:

$$(n_1 \approx n) \wedge (n_1' \approx n') \wedge (\exists (n_1, n_1') \in B_1 | (n_1, n_1') \approx (n, n')).$$

4. Discussion and Experimental Results

A preliminary evaluation of the algorithm has been carried out by considering an exact isomorphism problem, without taking into account semantic information. It is worth pointing out that semantic information is dependent on the considered application domain and its use generally allows to significantly reduce the search space. Consequently, the performance obtained by considering only the syntactic feasibility rules represents an upper bound of the matching time.

We have compared our method with the Ullman's algorithm [6] which is recognized among the most efficient ones. A set of connected graphs with a number N of nodes varying from 10 to 500, and an average number of branches equal to ηN^2, with $\eta \in [0.01, 0.03, 0.1, 0.25]$, have been randomly generated. Table 1 reports the matching times obtained implementing our algorithm in $C++$ under Linux, on a Pentium 120 MHz, with 32 Mb of RAM. As for the Ullman's algorithm, we used the $C++$ implementation reported in [9]. Note that the results of some tests with the Ullman's algorithm are not reported in Table 1 since memory requirements significantly increased with the number of nodes, overcoming the memory available on the adopted system.

In Fig. 6 we report, in both linear and logarithmic scale, the matching times versus number of nodes for $\eta = 0.03$. Both the algorithms show an exponential

Table 1. Matching times (in milliseconds) for different values of η. Column A: Results obtained with the implementation of Ullman's algorithm reported in [9]. Column B: Results obtained with our algorithm

Nodes	$\eta = 0.01$ A	$\eta = 0.01$ B	$\eta = 0.03$ A	$\eta = 0.03$ B	$\eta = 0.1$ A	$\eta = 0.1$ B	$\eta = 0.25$ A	$\eta = 0.25$ B
10	4.26	0.34	4.25	0.41	4.06	0.36	4.39	0.38
25	54.5	2.54	49.17	1.44	50.13	1.35	50.7	1.76
50	405.8	9.8	283.8	4.46	379.4	4.86	411.2	6.88
75	1409	37.9	1320.5	8.6	1263.5	10.8	–	16.5
100	3229	49.2	3039	15.4	3128	19.4	–	25.3
125	6176	56.7	6370	23.6	–	30.4	–	42.9
150	10285	65	10933	39.6	–	47.1	–	68.3
200	–	92.2	–	65	–	80.6	–	115.2
300	–	176.2	–	155.4	–	195	–	267.8
400	–	262.3	–	277.8	–	357.3	–	506.4
500	–	417	–	443	–	549.5	–	788.7

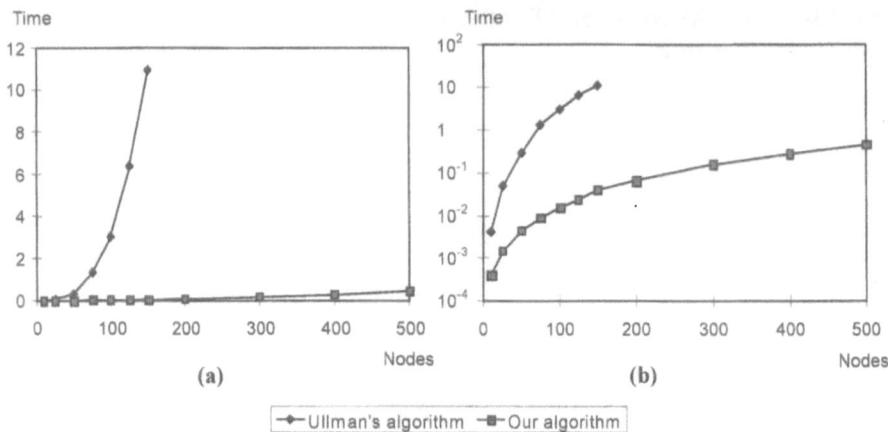

Figure 6. Matching time versus number of nodes for $\eta = 0.03$ with times expressed in seconds in linear scale **(a)** and logarithmic scale **(b)**

trend, but the coefficient of the exponential is lower for our algorithm. The absolute matching times are up to two orders of magnitude lower in our case.

Future investigations will be devoted to evaluate the performance of the method in case of the graph-subgraph isomorphism, and to characterize its computational complexity in case of both exact and inexact matching.

References

[1] Shapiro, L. G., Haralick, R. M.: Structural description and inexact matching. IEEE Trans. Pattern Anal. Mach. Intell. *3*, 505–519 (1981).
[2] Tsai, W. H., Fu, K. S.: Error-correcting isomorphisms of attributed relational graphs for pattern analysis. IEEE Trans. Syst. Man Cybern. *9*, 757–768 (1979).
[3] Tsai, W. H., Fu, K. S.: Subgraph error-correcting isomorphisms for syntactic pattern recognition. IEEE Trans. Syst. Man Cybern. *13*, 48–62 (1983).
[4] Sanfeliu, A., Fu, K. S.: A distance measure between attributed relational graphs for pattern recognition. IEEE Trans. Syst. Man Cybern. *13*, 353–362 (1983).
[5] Shapiro, L. G., Haralick, R. M.: A metric for comparing relational descriptions. IEEE Trans. Pattern Anal. Mach. Intell. *7*, 90–94 (1985).
[6] Ullman, J. R.: An algorithm for subgraph isomorphism. J. Assoc. Comput. Mach. *23*, 31–42 (1976).
[7] Bunke, H., Messmer, B. T.: Efficient attributed graph matching and its application to image analysis. In: Image analysis and processing (Braccini, C., DeFloriani, L., Vernazza, G., eds.), pp. 45–55. Berlin Heidelberg New York Tokyo: Springer 1995. (Lecture Notes in Computer Science, vol. 974)
[8] Cordella, L. P., Foggia, P., Sansone, C., Vento, M.: An efficient algorithm for the inexact matching of ARG graphs using a contextual transformational model. Proc. 13th ICPR. IEEE Comput. Society Press *3*, 180–184 (1996).
[9] Messmer, B. T.: Efficient graph matching algorithms for preprocessed model graphs. Ph.D. Thesis, Institute of Computer Science and Applied Mathematics, University of Bern, 1996.

L. P. Cordella
P. Foggia
C. Sansone
M. Vento
Dipartimento di Informatica e Sistemistica
Università di Napoli "Federico II"
Via Claudio 21, I-80125 Napoli, Italy
E-mail:
{cordel,foggiapa,carlosan,vento}@unina.it

Computing Suppl 12, 53–62 (1998)

© Springer-Verlag 1998

Efficient Graph Matching for Video Indexing

K. Shearer, Perth, **H. Bunke**, Bern, **S. Venkatesh**, and **D. Kieronska**, Perth

Abstract

Traditionally graph algorithms have been of restricted use due to their exponential computational complexity in the general case. Recently a new class of graph algorithms for subgraph isomorphism detection has been proposed, one of these algorithms having quadratic time complexity. These new algorithms use a preprocessing step to allow rapid matching of an input graph against a database of model graphs. We present a new algorithm for largest common subgraph detection that provides a significant performance improvement over previous algorithms. This new algorithm is based on the work on preprocessed subgraph isomorphism detection by Messmer and Bunke [3].

AMS Subject Classifications: 68Q20, 68T99.

Key words: Graph matching, image retrieval, video database, graph similarity, largest common subgraph.

1. Introduction

Graph algorithms have been used in many application areas as a formalism which supports the operations required. If a problem can be encoded as a graph, then a wide variety of well documented algorithms may be drawn upon, each of which has a known complexity. For examples of the application of graph algorithms in the field of pattern recognition see [2, 6, 8, 15]. The disadvantage with many graph algorithms is that they are exponential in complexity for the general case. Recent work by Messmer and Bunke [3] has introduced a new class of graph algorithms, which use a preprocessing step of exponential complexity to reduce the complexity of on-line processing.

The image correspondence problem is often cast as a graph isomorphism problem, with two images encoded as graphs, and a graph or subgraph isomorphism representing similar portions of the images. Graph matching algorithms also have potential in image and video database applications [4, 5, 9, 12]. In this area we have a database of perhaps thousands of images, against which we wish to perform queries of exact retrieval and similarity matching. The algorithms proposed by Messmer and Bunke are ideally suited to this application. These new algorithms allow us to preprocess the database off-line, then perform subgraph isomorphism detection in greatly reduced time. At worst the time complexity of the new algorithms is sublinear in the number of models in the database, and one algorithm is quadratic in the number of vertices in the input, that is independent of the number of models in the database. Given a large database this is a clear advantage over previous representations which require the comparison of each model with the input in a sequential manner, and may be exponential for each model.

2. Graph Encoding of Image and Video Information

There is a large body of work on representing images and image sequences using qualitative representations. The aim of these efforts is to provide a highly compact representation that allows rapid and intuitive retrieval of image information. The most successful form of representation of this type is the 2D-string [4, 7, 9]. A number of different 2D string representations have been proposed, which use a variety of operators. In essence each string representation uses a string of object identifiers to describe the relationships of objects in a picture along the u and v axes (horizontal and vertical axes). As an example the picture in Fig. 2a would be represented in 2D B-string notation [7] as: $(abBAcCdD, adAcD = bCB)$. Here a lower case letter represents the start point of an object and an upper case letter the end point. This simple start and end point labeling is used only for illustration purposes.

The majority of 2D string representations perform matching based upon some form of symbol ranking and comparison. Matching is divided into three *types*, called type-0, type-1 and type-2, with increasing strictness of match. For example, 2D B-strings [7] use five *relationship categories*: disjoint, meets, overlaps, contains and inside, to define type-0 matching. A pair of objects o_i and o_j, is a type-0 matching pair between two images A and B, if the relationships r_A and r_B, where $o_i r_A o_j$ in A and $o_i r_B o_j$ in B, are members of the same relationship category. For example in Fig. 1, all pictures are a type-0 match, as the relationships between the objects fall into the disjoint category in each case. Type-1 matching is made more strict than type-0 by ensuring that not only are two objects a type-0 match, but that they also display the same *orthogonal relationships*. Orthogonal relationships are whether an object is north (above), south (below), east (to the right of) and west (to the left of) of another object. In Fig. 1, pictures 2 and 3 are type-1 matches for picture q, both having B north and east of A, and A south and west of B, however picture 1 is not. Type-0 can be seen as an attempt at a scale and rotation invariant matching method, with type-1 being a restriction to cases which have the same orientation.

Type-2 matching requires the same actual relationship between the objects along each axis. Actual relationships are derived from work by Allen in temporal logic [1]. There are 13 possible relations between two intervals in one dimension, these relationships and the symbols representing them are presented in Table 1. Applying these relations along the two axis in two dimensions gives 169 possible relationships between two objects. An example graph encoding of two pictures can be seen in Fig. 2. The relationships in picture q are: A overlaps B along the hori-

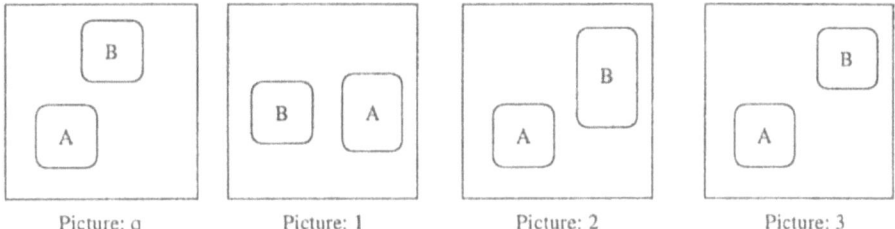

Figure 1. Example of type-n matching

Table 1. Possible interval relationships

Relation	Symbol	Example	Relation	Symbol	Example
less than	a<b		meets	a|b	
overlaps	a/b		ends	a]b	
contains	a%b		begins	a[b	
equals	a=b		begins inverse	a['b	
contains inverse	a%'b		ends inverse	a]'b	
overlaps inverse	a/'b		meets inverse	a|'b	
less than inverse	a<'b				

zontal axis, and A is less then B along the vertical axis. Picture q is thus not a type-2 match for any of the other pictures as they display the following relationships:

Picture	Horizontal	Vertical
1	B is less than A	B is contained within A
2	A is less than B	A overlaps B
3	A is less than B	A is less than B

These three types of matching allow queries over pictorial information with varying degrees of accuracy. A query may search for images with objects arranged with relationships identically or with one of two levels of approximation. This aids browsing as it allows relaxation or tightening of constraints for navigation of a database. In addition to inexact relationships in matching it is also important to

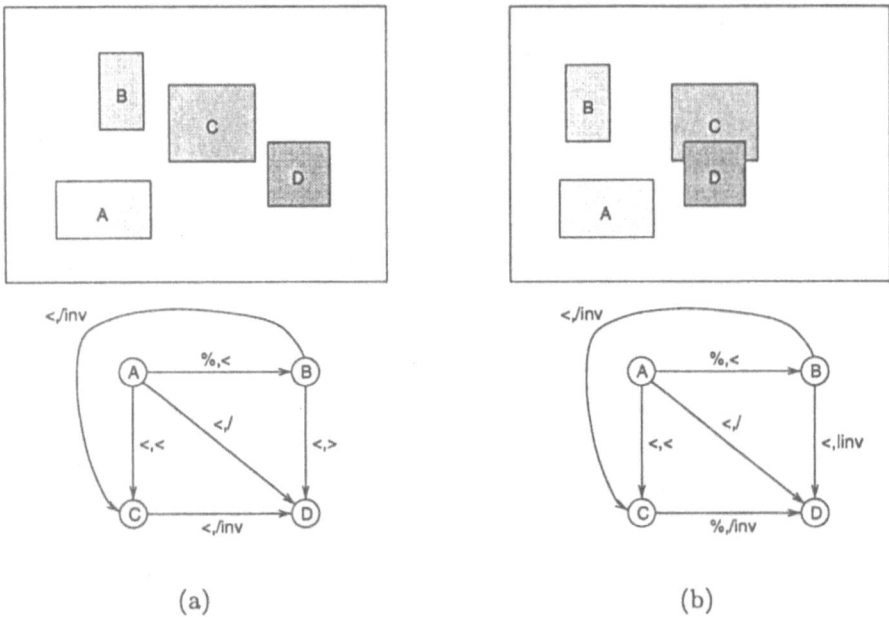

(a)   (b)

Figure 2a, b. Examples for similarity

consider cases where the object set present in images may not be identical. This may arise from searches for subpicture elements, or perhaps when browsing a database for images similar to an example image. If there is no picture with identical objects and relationships, it is often desirable to search for the picture that is most similar.

Given that the matching scheme is defined for object pairs, a complete pictorial match occurs if

$$\forall i, j : o_i, o_j \in A \Leftarrow \exists o_i, o_j \in B \text{ and } r_A \equiv r_B \text{ where } o_i r_A o_j \text{ in } A \text{ and } o_i r_B o_j \text{ in } A$$

Searching for the best inexact match in a pictorial database is termed *similarity retrieval*.

The most common similarity measure for 2D strings is based on the association graph formed by taking each pictorial object as a vertex, and inserting an edge between two vertices if they are a type-n match. Figure 3 shows the association graph for the pictures in Fig. 2. The similarity measure for two pictures is then the largest clique in the association graph. This largest clique is equivalent to the largest common subgraph of the two graphs and represents the largest collection of objects found in two images that maintain the same relationship (under the chosen matching type) to each other in both images. Figure 2 shows two simple pictures and their graph encodings. The main difference between the two pictures is that object D is higher and further left in Fig. 2b. The association graph in Fig. 3, constructed for type-2 matching, indicates that the largest common subgraph is

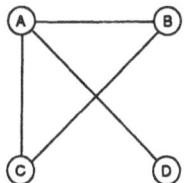

Figure 3. Association graph

ABC. This is the largest section of the pictures which appears the same in two pictures being compared.

3. Largest Common Subgraph Matching for Image Retrieval

In earlier systems for image retrieval from databases using spatial information, the maximal clique detection algorithm proposed by Levi [10] was used for finding the largest common subgraph. This algorithm first builds an association graph G_A, for the two graphs to be processed, then finds the maximal cliques in G_A. For an input graph of n vertices, and a database of L models of size m vertices, where we assume $n < m$, the clique matching algorithm has complexity of $O(L(nm)^n)$ in the worst case, and $O(Lnm)$ in the best case. The best case is two graphs with each node uniquely labeled such that no cliques of size greater than one exist, however in general complexity will be considerably greater than the best case.

This performance means that classical graph algorithms are often unsuitable for pictorial database work as the response time is unacceptable. The decision tree algorithm for subgraph isomorphism detection proposed in [3] has complexity which is polynomial in the number of nodes in the input graph. That is, it is independent of the number of models in the database, and the size of these models. This is a great reduction in complexity. As this algorithm is intended for finding subgraphs of the models isomorphic to the input graph this is clearly a large gain in speed.

While the decision tree algorithm provides a rapid method for finding subgraph isomorphisms from the input to the models, there is no version of the algorithm that allows similarity retrieval.

In this paper we suggest a new algorithm, based on the polynomial decision tree matching algorithm, to find the largest common subgraph of an input graph and a set of model graphs. This largest common subgraph may exist in one, or many of the model graphs, and the complexity is independent of the number of models, and also of the size of the models.

The original decision tree matching algorithm represents each graph by its adjacency matrix. Figure 4 gives an example of this; each entry (i,j) in the matrix is set to the label of the ith vertex if $i = j$, or the label of the edge between the ith and j^{th} vertex if $i \neq j$. A matrix entry of zero indicates that no edge is present between

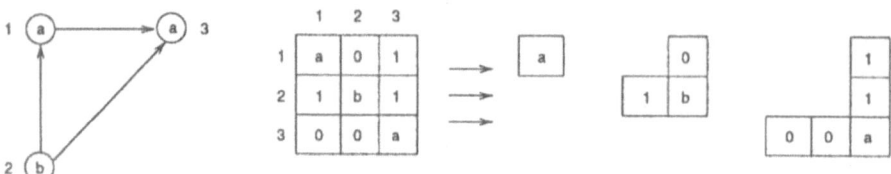

Figure 4. Graph with adjacency matrix and row column elements

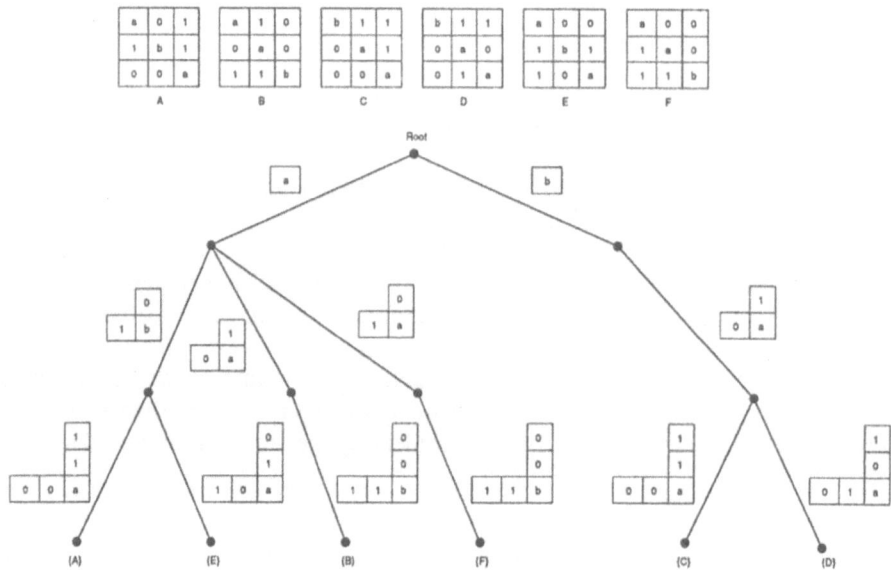

Figure 5. Decision tree for example graph

the vertices. The adjacency matrix may then be broken into row-column elements
as shown in Fig. 4. Each adjacency matrix which represents a model graph is
compiled into a decision tree, with each arc labeled with a row-column element of
the adjacency matrix. The decision tree for the graph in Fig. 4 is given in Fig. 5.
An input graph is then classified by comparing successive row-column elements
from its adjacency matrix with the arc labels of the decision tree. The algorithm
terminates when either the last row-column element of the input is matched, or
there is no match for a row-column element from the input. When the last element is matched we have detected a graph, or subgraph, isomorphism between
the input and at least one model graph represented by the tree. This isomorphism
is retrieved from the node of the decision tree at the end of arc which matched the
last row-column element. If there is no arc matching the next input row-column
element, then the algorithm exists as there is no graph or subgraph isomorphism
from the input to any of the models.

The initial observation that leads to the largest common subgraph (LCSG) algorithm is that at the point when the decision tree algorithm fails, there may in fact be further row-column elements in the input which can be matched against the decision tree. These may be accessed by permuting the row-column element which does not match to the bottom of the adjacency matrix, making note of this permutation, then continuing with the classification. This process terminates at the point where the row-column elements below the current point have all been tried and fail. At this point we have a candidate for the largest common subgraph, being the subgraph isomorphism represented by the partial adjacency matrix which was matched. However, it must be noted that an early choice of vertices may have led to the detection of a local maxima, so it is necessary to backtrack. We note that at a point where $e_1 \ldots e_m$ are matching row column elements, and $e_{m+1} \ldots e_n$ are the failures, we can backtrack by permuting the input adjacency matrix to be $e_1 \ldots e_{m-1} e_{m+1} \ldots e_n e_m$, and reducing the dimension of the matrix by one. Thus the combinations of $e_1 \ldots e_m$ are not examined again, as the last row-column element is no examined. This is sufficient as all combinations with $e_1 \ldots e_m$ have already been tested. If we also keep a record of the order of the largest common subgraph found, we can prune the search space when the best order, added to the number of permutations required to reach the current state, is greater than the current working dimension of the input adjacency matrix. At this point the order of the largest common subgraph so far plus the number of vertices excluded exceeds the order of the graph being tested, so there can be no improvement. It can be clearly seen that if the largest common subgraph is a large portion of the input graph, very little of the search space will be examined.

The best case computational complexity of this algorithm is $O(n^2)$, which occurs when the input is an exact subgraph of at least one of the models. Worst case complexity is difficult to define. To produce an estimate for the worst case complexity we use a result due to Turan [13].

Theorem 1 Turan. *Every graph G on n vertices with*

$$1 + \binom{n}{2} - t \cdot \frac{n - c - r}{2}$$

edges contains a clique of order $c + 1$, where $n = tc + r, 0 \le r < c$. This is the best possible.

The only graph G' of size

$$\binom{n}{2} - t \cdot \frac{n - c - r}{2}$$

which does not have a clique of order $c + 1$ is the complete c-partite graph with r parts of order $t + 1$ and $c - r$ parts of order t.

The graph G' gives an indication of the maximal number of order c cliques possible in a graph of order n. Using this indication we can estimate that the worst case complexity occurs when the largest common subgraph is approximately half the size of the input, and there is a maximal number of common subgraphs of order one less than the largest common subgraph. In this case the computational complexity is $O(2^n)$. This compares favourably with the clique matching algorithms worst case of $O(L(nm)^n)$. It should be stressed that we consider the worst case complexity to be an overestimate, and experimental results support this theory.

4. Experimental Results

We have conducted experiments using a database of videos mainly containing people walking in various surroundings. Much of the video is from a database used for a campus guide system [11]. There are three experiments reported here examining different aspects of the new algorithm.

The video used was sampled at a rate of approximately 5 frames per second, and each frame being represented by a model graph. The database is sparse in that within each clip, frames which produce identical models are represented by a single model in the database. If frames from separate clips produce identical models, the model will be added to the database twice. This resulted in a maximum of 562 models being added to the database.

In the first experiment there are two groups of results. Group one consists of four queries containing four vertices each. This is considered a typical size for an initial query by example retrieval. These queries were performed on a database of video clips representing 7 minutes and 41 seconds of filming. The second group of queries gives one query of 6 vertices and two further queries of 7 vertices. These queries were performed on a database of video representing 6 minutes and 30 seconds of filming. There are between 9 and 12 annotated objects in each clip, each is represented by a vertex in the graph representation.

Each set of queries was performed using the new largest common subgraph algorithm (LCSG) as described in Section 3, the best exact subgraph isomorphism method (ESI) and a similarity measure based on graph edit distance (GED) [3]. Ullman's algorithm is used as the ESI [14], and A^* with lookahead as the GED. The times reported are in milliseconds, when run on a Silicon Graphics Iris, with a R4400 processor and 96Mb of main memory. The results of the experiments are presented in Table 2.

In the second experiment the full database of clips was loaded, giving approximately 9 minutes 22 seconds of video. Four queries where then run for the three inexact algorithms: A^*, inexact network and LCSG. The results from this experiment are presented in Table 3. The first two queries are frames randomly chosen from the video database. These test the performance of the algorithms when an exact match is available. The third and forth queries show the performance of the algorithms as the error in the input graph increases. These queries show a similar level of performance difference to experiment one for the exact retrieval, but a greatly increased difference as the error in the input increases.

Table 2. Experimental results

Group 1	LCSG	ESI	GED	Group 2	LCSG	ESI	GED
query 1	8	189	3800	query 1	16	236	19773
query 2	18	199	7161	query 2	16	249	22207
query 3	7	172	3810	query 3	18	251	21336
query 4	11	202	2306				

Table 3. Inexact isomorphism detection

A* (GED)			Inexact network			New LCSG		
Query	Error	Time	Query	Error	Time	Query	|LCSG|	Time
Liblr.10	0	10737	liblr.10	0	172	liblr.10	10	26
liblr.0	0	9073	liblr.0	0	164	liblr.0	9	22
wayq	6	9122	wayq	6	223	wayq	4	7
libq	12	35674	libq	12	2851	libq	6	14

Table 4. Approximate match against 11 graphs

Query	A* (GED)	Network	LCSG
11.6	282	30	17
12.2	53	15	5
14.1	84	18	7
15.2	642	34	23
15.3	692	36	22
16.1	136	20	8

The final experiment gives the results of queries run against a set of eleven graphs, with an average of 8 vertices. The queries used were deliberately created to extract the worst performance from the LCSG algorithm. As can be seen in Table 4, even for the worst case of the LCSG over a small number of models, the new algorithm is considerably faster than previous algorithms.

The experiments clearly show the advantage of dissociating the computational complexity from the number of models and the size of the models. Using the new algorithm gives at least an order of magnitude better time than the exact algorithm in all cases. The improvement over traditional similarity methods achieved by the new algorithm indicates the suitability of this work for the image and video database area.

5. Conclusion

We introduce a new algorithm for finding the largest common subgraph between a database of model graphs and an input graph given on line. This algorithm is based on the preprocessed graph matching concepts of Messmer and Bunke. The new algorithm, with worst case complexity $O(2^n)$, greatly improves upon the previous best algorithm for largest common subgraph detection which has worst case complexity $O((nm)^n)$. The complexity of the new algorithm is also independent of the number of models in the database, which the previous algorithm is not. This provides a similarity measure for graphs which is rapidly computable, and in some cases more appropriate than edit distance. For example, image similarity using spatial representations is best measured using the largest common subgraph.

It should be noted, however, that the algorithm requires an exponential complexity preprocessing step, and exponential space. This limits the applicability of the work to applications which have limited numbers of vertices in their graph representation. Fortunately there are numerous such applications in the domain of image and video databases. Further work is being done to address the space complexity problem.

References

[1] Allen, J. F.: Maintaining knowledge about temporal intervals. Comm. ACM 26, 832–843 (1983).
[2] Bunke, H., Allerman, G.: Inexact graph matching for structural pattern recognition. Pattern Rec. Lett. 1, 245–253 (1983).
[3] Bunke, H., Messmer, B.: Recent advances in graph matching. International J. Pattern Rec. Art. Intell. 11, 169–203 (1997)
[4] Chang, S., Shi, Q., Yan C.: Iconic indexing by 2D strings. In: Proceedings of the IEEE Workshop on visual languages, Dallas, Texas, USA, June 1986. Also in: IEEE Trans. Pattern Anal. Mach. Intell. 9, 413–428 (1987).
[5] Chang, S. K., Yan, C. W., Dimitroff, D. C., Arndt, T.: An intelligent image database system. IEEE Trans. Software Eng. 14, 681–688 (1988).
[6] Horaud, R., Skordas, T.: Stereo correspondance through feature grouping and maximal cliques. IEEE Trans. Pattern Anal. Mach. Intell. 11, 1168–1180 (1989).
[7] Lee, S., Yang, M., Chen, J.: Signature file as a spatial filter for iconic image database. J. Visual Lang. Comput. 3, 373–397 (1992).
[8] Lee, S. W., Ren, Y., Suen, C. Y.: Hierarchical attributed graph representation and recognition of chinese characters. Pattern Rec. 24, 617–632 (1991).
[9] Lee, S.-Y., Hsu, F.-J.: Spatial reasoning and similarity retrieval of images using 2D C-string knowledge representation. Pattern Rec. 25, 305–318 (1992).
[10] Levi, G.: A note on the derivation of maximal common subgraphs of two directed or undirected graphs. Calcolo 9, 341–354 (1972).
[11] Shearer, K. R., Venkatesh, S., Kieronska, D.: Spatial indexing for video databases. J. Visual Comm. Image Repr. 8, 325–335 (1997).
[12] Tamura, H., Yokoya, N.: Image database systems: A survey. Pattern Rec. 17, 29–43 (1984).
[13] Turán, P.: Eine extremalaufgabe aus der graphentheorie. Mat. Fiz. Lapok 48, 436–452 (1941).
[14] Ullman, J. R.: An algorithm for subgraph isomorphism. J Assoc. Comput. Mach. 23, 31–42 (1976).
[15] Wong, E. K.: Model matching in robot vision by subgraph isomorphism. Pattern Rec. 25, 287–304 (1992).

K. Shearer
S. Venkatesh
D. Kieronska
Department of Computer Science
Curtin University of Technology Perth
Australia
e-mail: kims@cs.curtin.edu.au

H. Bunke
Institut für Informatik und Angewandte
Mathematik
Universität Bern
e-mail: bunke@iam.unibe.ch

Computing Suppl 12, 63 – 71 (1998)

Isomorphism between Strong Fuzzy Relational Graphs Based on *k*-Formulae

L. Wendling and **J. Desachy,** Toulouse

Abstract

We present a new graph matching approach based on 1D information. Each node of the graphs represents a fuzzy region (fuzzy segmentation step). Each couple of nodes is linked by a relational histogram which can be assumed to the attraction of two regions following a set of directions. The attraction is computed by a continuous function, depending on the distance of the confronted objects. Each case of the histogram corresponds to a particular direction. Then, relational graph computed from strong scenes are matched.

Key words: Pattern recognition, image processing, discrete algorithms matching, fuzzy relations, graphs.

1. Introduction

In this paper a new pattern recognition system, based on sample images, is presented. An object to be recognized is described by one scene or by a set of little images. Then, a fuzzy segmentation step is performed in order to split an image into a fuzzy partition which consists on a set of fuzzy regions. For each fuzzy region both topological and relational features can be computed (methods based on atomic regions and hierarchical trees have also been proposed in previous papers [6], [21]). In the proposed approach each couple of regions of the image is linked by a histogram of forces. So, an image is defined as a relational graph. The same process is applied to another images to match. Then, a new approach of matching based on strong relational graph is defined.

2. Pattern Recognition System

Our pattern recognition system (Fig. 1) consists in five parts.

1. Input data (grey levels or RGB color images).
2. Fuzzy segmentation: To split the image into a set of fuzzy regions.
3. Relations to be computed.
4. A data base composed of typical relational graphs to match.
5. Decision part to give a distance measure between two matched scenes.

Each part performs a particular process to split the image into a relational graph and to take a decision: Scenes are similar or not.

L. Wendling and J. Desachy

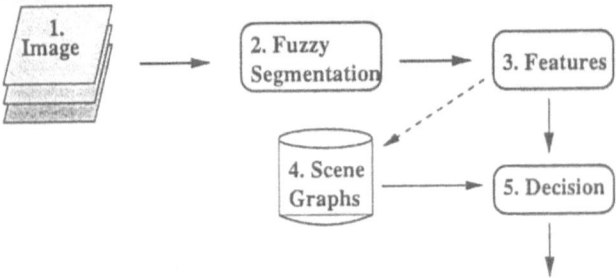

Figure 1. System description

3. Fuzzy Region Definition

First part of this system performs a fuzzy segmentation from the input image. The goal of fuzzy segmentation method is to manage with imprecise boundaries [19] [10] and to allow to a pixel to belong more or less to a given region. Crisp segmentation [14] is a particular case of fuzzy segmentation and can be achieved with a max criteria applied to fuzzy partitions [3]. Most of the fuzzy segmentation methods are based on the definition of fuzzy partitions using fuzzy c-means algorithms [3]. Nevertheless these approaches give noisy and non-totally coherent results [9] [4].

A recent method proposed by Krishnapuram [11] seems to overcome these problems. His method is independent of the interclass distance and is based on a "good" membership profile [22]. The initialisation of the defined algorithm is fundamental to achieve partitions. Barny et al. [1] have shown that the use of c-means algorithm to define the input partition can fail by defining indentical clusters. This problem can be overcome if possiblilistic c-means algorithm [13] is used with a number of classes equal to 1. First any partition point is set at 1. Then the most favourable cluster (defined from both validity criteria and partitions variations between two steps) is carried out. If such a class is found, points of cluster data which most verify this achieved cluster are removed from the image partition (clusters RGB in a color image, for example). Processing is runned again until the achievement of unconsistent clusters (too small for example) is performed. Currently clusters validity algorithms have generally high processing time with sometimes unconsistent results [9]. Moreover, it exists no mathematic models to define what is a "good" partition.

In the present system, this algorithm has been applied with a level cut criteria to decrease processing time (0.8s on a 100 MHz SUN SPARC 4). Then a partition composed of fuzzy clusters is achieved.

A fuzzy region is defined as being a set of connected pixels with a non-zero membership value. At this step, the system has defined the set of fuzzy regions (nodes of the graph). Then, we can link these regions with relational attributes to define a relational graph.

4. Fuzzy Relational Graph

First, each region is assumed to its centroid. Then, we can compute the 1D relation which links each couple of nodes. In previous works, features, computed on pixels [17] or level-cuts [5] [8], are often used to distinguish objects. The main problem of these methods is to define the most significant features which depend on the application. In this paper, a new relational matching based on histogram of angles and forces notion is proposed. Histogram of forces is a generalization of Miyajima and Ralescu histogram of angles [16] with isotrop segments. Let A and B be two objects and θ a direction. The histogram value corresponding to θ consists in a Riemann sum of combination of segments of A and B. A object is composed by a set of parallel segments (one pixel height) following a direction. The function apply between a segment of a region A and a segment of a region B, bear by the same straight line, takes into account the distance notion: The farer the objects, the lesser the value of the linking continuous function is. This can be assumed to be the projection of the information onto an one dimension space in regard to both matched regions. This approach allows to have a low processing time and to manage with non-disjoint and fuzzy objects. Such a function is entirely detailed in [15].

5. Matching between Strong Graphs

5.1. Graph Structure

It is well known that graph isomorphism problem is not deterministic (except for special kinds of graphs). The present approach manages with a particular class of graphs: Graphs with strong structure. The structure is achieved after computing histogram of forces between regions using their centroid).

In a graph G, an edge between two nodes s and t is single. A set of values is beared by each edge. Edges can be double, but as information, given by histogram of forces, is symmetric a preorder is fixed.

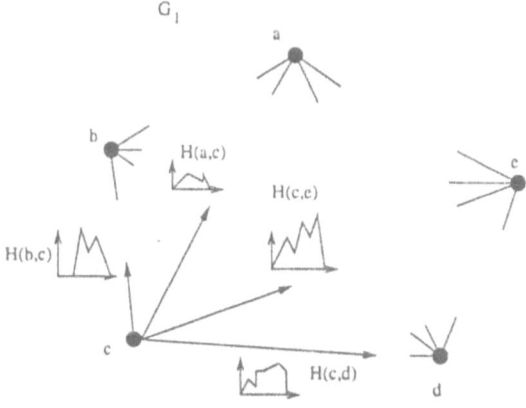

Figure 2. Relational graph

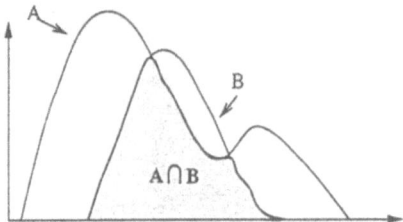

Figure 3. Histogram superimposition

5.2. Similarity Ratio

Let A be a histogram of forces linking vertex s to t in a graph G_1 and B be the histogram of forces linking vertex u to v in a graph G_2. A and B superimposed in order to compute a distance between them from their common parts (Fig. 3).

In the present application, a similarity ratio has been chosen to match A and B. It is obvious that other distance measures can be used. Nevertheless, the calculus of a similarity ratio is low processing time and takes into account all the histogram information. Let v be the number of steps of each histogram, i.e. the number of digitized directions. The cardinal of intersection of histogram A and B is given by:

$$|A \cap B| = \sum_{i=1,v} \min(A[i], B[i])$$

$|A|$ represents the cardinal of the histogram A. By definition, A and B can be null if the nodes (fuzzy regions) s, t, u and v exist (this is due to the positive function applied). Then, the similarity ratio is computed as follows:

$$S(A, B) = \frac{|A \cap B|}{\max(|A|, |B|)}$$

5.3. k-Formula Computation

Strong graphs, i.e. rigid structures, are taken into account. Let θ be an orientation and let s and t be two nodes of a graph G_1. Let P_t and P_s be the respective projections of s and t following a directional straight line D_θ defined with a θ angle rotation from the frame image. If P_t is lower than P_s on D_θ (Fig. 4b), an edge from t to s, denoted $t \rightarrow_{G_1} s$, is carried out. Vertice are sorted following θ orientation to built the associated k-formula. This computing is performed for all the nodes of the Graph G_1. The set of k-formulas which composed the graph, following the θ direction, is so defined. The same processing is carried out for all the nodes of the graph G_2 and following a direction. Then, k-formulas [2] have been generated from each graph (Fig. 4a). At last, the k-formulas of graph G_1 are matched with the K-formulas of graph G_2.

The final recognition rate is given either by the mean similarity ratio following the optimal direction (which corresponds to a histogram shift) or by the minimum

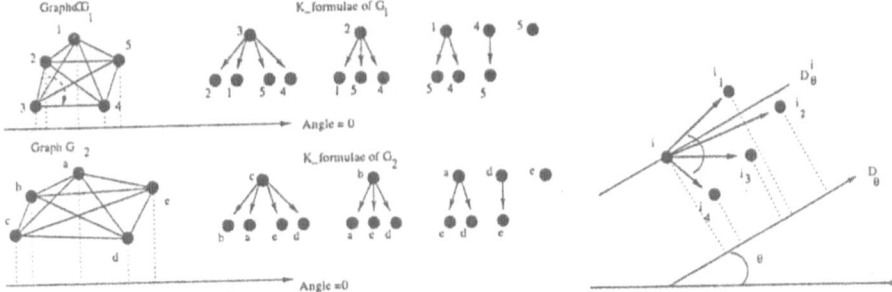

Figure 4. k-formulae definition (projection)

similarity ratio following the optimal direction. The choice of the decision criteria (minimal or mean) can be brought by the number of nodes and depends to the kind of the current application. For example, to take a minimum criteria with graph belonging a large number of nodes can induce a drowning phenomena. An application, of such an approach, to the case of real scenes matching is presented in Section 6.

5.4. Matching Algorithm

The processing performed by our pattern recognition system scheme can be summarized as follows:

Match (G_1, G_2)
Input: Two Graphs.
Output: Recognition Rate.
{
   Fuzzy Segmentation
   Regions Localization and Graph building
   Histograms of forces of G_1 and G_2
   k-formulas definition of G_1 (quicksort)
   Similarity $= 0$
   For any direction θ (histogram digitization) Do
   {
     k-formulas definition of G_2 (quicksort)
     $\lambda \Leftarrow$ Matching between k-formulas of G_1 and G_2
     Similarity $\Leftarrow$ max(Similarity, λ)
   }
}

Figure 5. General Algorithm

5.5 Complexity

The case of any k-formula for a given graph is processed. Given an angle θ, comparisons between k-formulae of graphs G_1 and G_2 are carried out. This processing is performed for any θ and the recognition rate is set to the maximal similarity ratio. The maximal complexity associated to the matching is in $\mathcal{O}(n^2)$ time (with $n = |G_1| = |G_2|$).

Nevertheless complexity of the method depends on the k-formula definition. If a quicksort is used, a K-formula with n_1 sons is build in $\mathcal{O}(n_1 \ln n_1)$ time. So, the building of all the k-formulas of a graph G, with $|G| = n + 1$ needs $f = \sum_{i=1}^{n} i \ln i$ operations. Then, let us put down the following proposition:

Proposition 1. $\forall n \in N^\star$, we have:

$$\frac{1}{2}\left(n^2 \ln n - \frac{n^2}{2} - \frac{1}{2}\right) \leq \sum_{i=1}^{n} i \ln i \leq \frac{1}{2}\left((n+1)^2 \ln(n+1) - \frac{(n+1)^2}{2} - \frac{1}{2}\right)$$

A simple demonstration by recurrence can be used to verify these two inequalities. The function f is strictly increasing (and continuous). The boundaries functions have been defined by the integration of f following the rectangle method (area of f minored and majored by both minimal and maximal rectangle functions).

Finally, it is easy to deduce that the maximal complexity of the proposed approach is in $\mathcal{O}(n^2 \ln n)$ time.

6. Application with Color Scenes

An application of the previous described method is given now. It consists in matching two RGB scenes supposed strong. Both acquisition height and orientation of images are different. Two methods have been applied. First, an isomorphism search between graph G_1 (defined by the five regions included in clusters A_2 and A_3) and graph G_2 (clusters B_2 and B_3) is performed. For each couple of regions a histogram of forces is computed. The discretisation step is equal to $1/256$. Using a lower step, value variation of the final rate is under 10^{-4} order.

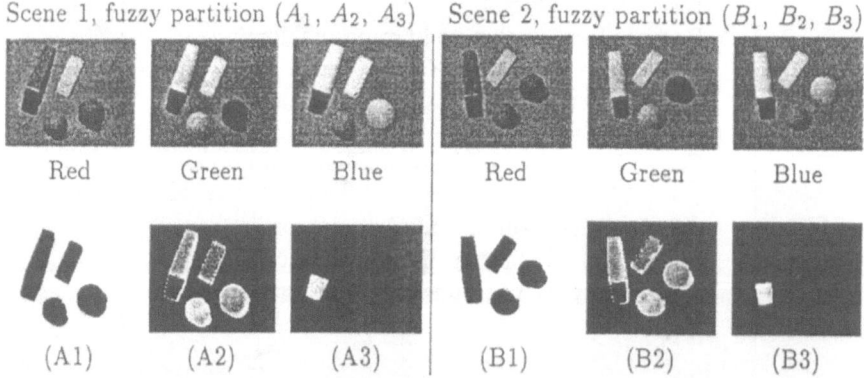

Scene 1, fuzzy partition (A_1, A_2, A_3) Scene 2, fuzzy partition (B_1, B_2, B_3)

Red   Green   Blue   Red   Green   Blue

(A1)   (A2)   (A3)   (B1)   (B2)   (B3)

Figure 6. Matched scenes

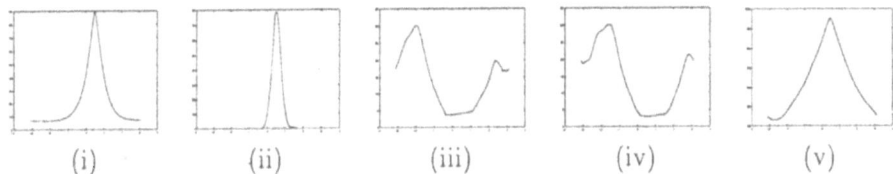

(i)   (ii)   (iii)   (iv)   (v)

Figure 7. (i), (ii), (v) Similarity ratio variation; (iii), (iv), (v) Histogram of forces between clusters

Figures 7(i) and 7(ii) show similarity ratio variation from angle rotation in $[-\pi, \pi]$ interval. A mean similarity rate of 89.84%, between the two scenes, is reached with a 28.13 degrees shift (case (i)). Of course, if the number of regions is important then similarity ratio should be high even if a region is bad. So, it can be interesting to compute the minimal similarity ratio (following optimal angle) (case (ii)). In the present example, it reached 78.47%.

These curves (i) and (ii) show that an A^* heuristic can be useful to decrease processing time. In the present approach an A^* heuristic has been defined to find the most characteristic directed acyclic graph. This algorithm is based on distance minimal from an edge of graph G_1 and any of graph G_2 (up to a translation in the matched histograms of forces to find the best probable direction) and by building G_2 with an optimal cost. The application of such a heuristic gives the same similarity ratio achieved result, but fastly. Moreover, this heuristic can be applied to the case of strong subgraphs.

In a second approach, each cluster is assumed to be a single region. The histogram of forces linking the second and the third cluster on Fig. 7 (iii) – A_2 with A_3 – and (iv) – B_2 with B_3 –. In this case, a maximal similarity ratio of 95.44% is reached for a 25.31 degrees shift. The acquisition image height is fairly different and histogram normalisation (to avoid the zoom factor) has given a weak improvement of 0.3%. Similarity ratio variations are given in Fig. 7 (v). This other approach is useful because it is not necessary to define each region (set of connected pixels with no zero membership value) contained in the clusters of a fuzzy partition. As a consequence, the number of matches is lower than in the previous method (ten time lower). This approach can be assumed to be a distance measure between two models.

Hence this method is less discriminant than the previous one which takes into account any region of a partition. Then, relational information are lost. A better ratio is reached but it does not give a better idea of the reality. This result is relative to the partition quality. In the present approach the main information is located in the second cluster (B_2 and A_2).

If a partition validity criteria is taken, which limits the number of regions per cluster and selects only the most dense part of the cluster, it is obvious to think that the number of clusters per partition should be increased and the number of regions per partition should be decreased. In this case, the second approach becomes more interesting because relational information is rather located in an inter-partition scheme. It is possible to apply this kind of approach even with noised areas (the function used to defined histogram of forces take into account disjoint information). Moreover, in the previous approach, noise must be removed to keep "dense" areas, which induces lost of information.

The proposed approach has been applied with success on more complex scenes (up to twenty five nodes and seven clusters). When we take into account two small regions the present approach fail down. That is not the case when we work only with clusters. Currently, we envisage to consider the search of maximal subgraph to improve the matching.

7. Conclusion

In this paper, a method of pattern recognition based on relational graphs and k-formulae definition has been proposed. This approach, which has a low processing time, has given interesting results using objects with strong structure. Currently, the present algorithm is applied to more complex scenes (about a hundrend of nodes) and we try to extend the present approach to the case of subgraph matching in order to take into account the problem of small regions.

Nevertheless, the present approach is limited on a particular class of graphs. We try to generalize our approach to manage with non-totally ordonned graph. Our aim is to define a progressive algorithm managing with the case of strong structure to general structure.

References

[1] Barni, M., Cappellini, V., Mecocci, A.: Comments on "A Possibilistic Approach to Clustering". IEEE Trans. Fuzzy Systems *4*, 393–396 (1996).
[2] Berztiss, A. T.: A backtrack procedure for isomorphism of directed graphs. Assoc. Comput. Mach. *20*, 365–377 (1973).
[3] Bezdek, J. C.: Pattern recognition with fuzzy objective function algorithms. New York: Plenum Press, 1981.
[4] Dave, R. N.: Characterization and detection of noise in clustering. Pattern Rec. Lett. *12*, 657–664 (1991).
[5] Dubois, D., Jaulent, M. C.: *A* general approach to parameter evaluation in fuzzy digital pictures. Pattern Rec. Lett. *6*, 251–261 (1987).
[6] Dhérété, P., Wendling, L., Desachy, J.: Fuzzy segmentation and astronomical images interpretation. ICIAP'95-IAPR, Lecture Notes in Computer Science, vol. 974, pp 21–27. Berlin Heidelberg New York Tokyo: Springer 1995.
[7] Jantzen, J., Ring, P.: Image segmentation based on scaled membership functions. 2nd IEEE Conference on Fuzzy Systems 2, pp. 714–718 (1993).
[8] Krishnapuram, R., Keller, J. M., Ma, Y.: Quantitative analysis of properties and spatial relations of fuzzy images regions, IEEE Trans. Fuzzy System *1*, 222–233 (1993).
[9] Krishnapuram, R.: Generation of membership functions via possibilistic clustering. 3rd IEEE Conf. Fuzzy System *3*, 902–908 (1994).
[10] Keller, J. M., Carpenter, C. L.: Image segmentation in the presence of uncertainty. Int. J. Intell. Syst. *5*, 193–208 (1990).
[11] Krishnapuram, R.: Fuzzy clustering methods in computer vision. EUFIT'93, pp. 720–730 (1993).
[12] Krishnapuram, R., Keller, J. M.: A possibilistic approach to clustering. IEEE Trans Fuzzy Syst. *1*, 98–110 (1993).

[13] Krishnapuram, R., Keller, J. M.: The possibilistic C-means algorithm: insights and recommendations. IEEE Trans. Fuzzy Systems 4, 385–393 (1996).
[14] Horowitz, S. L., Pavlidis, T.: Picture segmentation by a directed split and merge procedure. 2nd International Conference on Pattern Recognition, pp. 424–433 (1974).
[15] Matsakis, P., Wendling, L., Desachy, J.: Représentation de la position relative d'objets 2D au moyen d'un histogramme de forces. Rev. Trait. Signal 4, 63–76 (1998).
[16] Miyajima, K., Ralescu, A.: Spatial organization in 2D segmented images: representation and recognition of primitive spatial relations. Fuzzy Sets Systems 65, 225–236 (1994).
[17] Rosenfeld, A.: Fuzzy geometry: an overview. First IEEE Conf. Fuzzy Systems, pp. 113–118 (1992).
[18] Rosenfeld, A., Hummel, R., Zucker, S.: Scene labelling by relaxation operations. IEEE Trans. Syst. Man Cybernetics 6, 420–433 (1976).
[19] Ruspini, E. H.: A new approach to clustering. Inf. Control 15, 22–32 (1969).
[20] Wendling, L.: Segmentation floue appliquée à la reconnaissance d'objets dans les images numériques. Ph.D. thesis, Université Paul Sabatier, Toulouse, 1997.
[21] Wendling, L., Pariès, A., Desachy, J.: Pattern recognition by splitting images into trees of fuzzy regions. Int. J. Intell. Data Anal. 1(2) (1997). http://www-east.elsevier.com/ida
[22] H. J. Zimmerman, P. Zysno, Quantifying vagueness in decision models. Eur. J. Oper. Res. 22, 148–158 (1985).

L. Wendling
J. Desachy
Université Paul Sabatier – IRIT
118, route de Narbonne
F-31062 Toulouse Cedex, France
e-mail: wendling@irit.fr

Computing Suppl 12, 73 – 82 (1998)

A Graph Structure for Grey Value and Texture Segmentation

H. Jahn, Berlin

Abstract

An attempt for developing an unified method for grey value and texture segmentation was made. It makes use of a special graph structure (Feature Similarity Graph—FSG) which is based on a feature similarity criterion and a feature smoothing procedure applied in each layer of the network. Starting with grey value segmentation (the features are the pixel grey values) one obtains segments which, for textured images, represent texture elements (texels) or parts of texels and background, respectively. The texels can be described by certain features, namely position, orientation, size, grey value or color, and shape descriptors. Studying position and orientation, spatial frequency phenomena and important observations made by investigators of human perception, especially the Gestalt laws, can be explained. The highly parallel $O(N)$ method can be applied also to the clustering of dot patterns.

Key words: Segmentation, texture, edge detection, edge preserving smoothing, graphs.

1. Introduction

To cope with the basic problem of image segmentation various approaches have been developed [1], but there does not exist a general framework to deal with grey value *and* texture segmentation. Whereas grey value segmentation often is based on thresholding, region growing and related methods, for texture discrimination statistical methods, power spectra and special filters are used. Here an attempt is made to develop a method which allows an unique description of low level image processing operations such as segmentation, edge detection and edge preserving smoothing.

Hints for the development of such an unified method are given by the results of the investigations of pre-attentive vision and of graph structures for image description. Experiments carried out to investigate the pre-attentive texture segregation capabilities of humans [2, 3] showed that (1) there exist certain features of texture elements (texels), called textons, which local differences are responsible for, at least, parts of texture segmentation, (2) pre-attentive texture discrimination is carried out instantaneously and therefore must be done in parallel. It follows that at least a broad class of textures can be analysed using proper features and spatial arrangement (spatial frequency) of texels. This favors the so-called structural approach to texture [4] which, because of its difficulties, was not investigated deeply in the past. On the other side, the use of graphs for image description has proven to be adequate. Especially the adaptive and irregular pyramid based approaches to grey value segmentation [5, 6] demonstrate this. Because of necessary local, regional and global processing of image data a layered architecture as in pyramids [7] seems to be adequate. Inside a layer the processing should be massively parallel, and the information flows from the bottom (pixel level) to the top (segments).

The processing starts with grey value segmentation in the lower layers of the network. Result of that procedure are certain segments which may be objects, parts of objects or merges of objects. If there are many of these segments in the image and if they show a certain similarity and a more or less periodical arrangement, then these segments can be used as texels for texture segmentation in the higher layers of the network. In each grey value segmentation layer the pixels are the nodes of a special graph which, analogously to the RAG, can be called a Pixel Adjacency Graph (RAG) [8].

According to their similarity of grey value the nodes (or vertices) can be connected by edges (or branches). In each layer segments as the *connected components* of the PAG can be defined. Averaging over the grey values inside segments or parts of them guarantees noise reduction and proceeding from local to global processing. In a texture segmentation layer of the network the texels are the nodes of an appropriate graph. Each texel is described by a set of features. The texels can be connected by graph edges if their features show a certain similarity. Therefore, such a graph can be called a Feature Similarity Graph (FSG), and again the texture segments can be defined as the connected components of the FSG [9].

The method fulfills the following requirements:

1. The algorithm is fully parallel in each network layer.
2. Bottom-up processing starting with the image pixels representing the receptor elements is used.
3. The method ensures efficient noise reduction by local, regional, and global processing.
4. Any number of segments are possible because this number is not known a priori. Furthermore, there are no restrictions concerning the shape of segments.
5. Smooth spatial changes of features inside a region do not lead to segregation.
6. The algorithm adapts to changing image statistics without using special knowledge of objects.
7. If N is the number of image pixels then the complexity of the algorithm is not greater than $O(N)$ in each network layer.

In Section 2 the method is developed in general. Section 3 then is devoted to grey value segmentation. The proximity of texel positions and the similarity of the texel orientation is dealt with in Section 4.

2. The Feature Similarity Based Graph Network

We consider M points $P_k = (x_k, y_k)$ $(k = 0, \ldots, M - 1)$ which can be the center of gravity positions of texels (for texture segmentation) or the pixels (i, j) (for grey value segmentation), respectively. Let f_k be a feature assigned to P_k. The number k of P_k defines the node k of a Feature Similarity Graph (FSG). The nodes can be connected by a graph edge if a similarity criterion

$$|f_k - f_{k'}| \leq \mathrm{Var}(f) \tag{1}$$

is fulfilled. Here, the points P_k and $P_{k'}$ are spatially neighbored, and $\mathrm{Var}(f)$ is a measure of the variation of the feature f in a certain neighborhood of P_k and $P_{k'}$. (1) reflects the fact that not the features themselves but their differences (feature contrast [3]) are responsible for segmentation. Furthermore, the separability of regions depends on the feature variation inside the regions (or the target *and* the background) which is represented by $\mathrm{Var}(f)$.

A possibility to define the FSG to be used here is the Node Adjacency List NAL_f [10]. If (1) is fulfilled then node k' will be put into $\mathrm{NAL}_f(k)$ and node k into $\mathrm{NAL}_f(k')$, respectively. Now, it is possible to label the connected components of the FSG using a graph traversal algorithm, e.g. Depth First Traversal [10]. Then $\mathrm{LABEL}(k)$ assignes an integer number, the number of the segment, to each node k.

Now we consider a certain window W_k centered in P_k. Then, averaging the values of all features f_k, which nodes $P_{k'}$ are inside W_k and belong to the same connected component as P_k, we obtain an averaged value $\langle f_k \rangle$ instead of f_k. This process can be applied recursively in order to smooth out noise:

Let us consider the recursion level (or network layer) l. Input to level l are the averaged values $f_k(l - 1)$. Now the criterion

$$|f_k(l - 1) - f_{k'}(l - 1)| \leq \mathrm{Var}(f, l) \tag{2}$$

is applied to the smoothed feature values $f_k(l - 1)$ (the threshold Var now can depend on level l). Averaging over connected nodes inside a window $W_k(l)$ results in new smoothed feature values $f_k(l)$ which are the output of layer l and the input of layer $l + 1$. This process is terminated at level $l_{\max}$. The node adjacency list NAL_f of layer $l_{\max}$ defines the final FSG with the segments as its connected components. The structure of the graph shows the connections and the disconnections of points P_k. The disconnections of nodes represent the edges between different regions or edges inside such regions, and the values $\langle f_k \rangle = f_k(l_{\max})$ are smoothed feature values which can be used as segment characteristics for further processing. $\mathrm{LABEL}(k) = m$ assignes to each point P_k a segment number m.

In the next chapters we will specialize the developed method to features of different kind.

3. Grey Value Segmentation

Grey value segmentation (or segmentation by luminance) as one of the first steps of image processing in the human visual system is carried out by retinal cells, whereas textural segmentation as a higher level processing step is done in the cortex. Therefore, it is natural to start with grey value segmentation. A first attempt to formulate grey value segmentation using the presented method was made in [8].

Now, the points P_k are the image pixels (i,j) $(i, j = 0, \ldots, N - 1)$ and the features are the grey values g_{ij}. To define the Pixel Adjacency Graph (PAG) as a special version of the FSG only 4-connected pixels are considered. Let (i,j) and (i_1,j_1) be 4-neighbored pixels. Then, in level l (level 0 is the original image with grey values $g_{i,j}$), the similarity of the smoothed grey values of these pixels is checked using the similarity criterion

$$|g_{i,j}(l - 1) - g_{i_1,j_1}(l - 1)| \leq F_{i,j;i_1,j_1}(l). \tag{3}$$

Here, F is a certain threshold which depends on the grey value variation within a certain neighborhood of the pixels (i,j) and (i_1,j_1). I have checked many possibilities such as $F \sim \sigma$ or $F \sim \sigma\sqrt{1 - r}$ where σ is the grey value standard deviation and r the correlation coefficient between $g_{i,j}$ and g_{i_1,j_1}. Many thresholds give good results but here I emphasize the threshold

$$F_{i,j;i_1,j_1}(l) = \text{MAX}\{t_1(l) \cdot \text{MIN}[\Delta g_{i,j}, \Delta g_{i_1,j_1}]; t_2(l)\}, \tag{4}$$

because it is closely related to the threshold used for dot pattern clustering and texture segmentation (see below). In (4) $\Delta g_{i,j}$ is the minimum of $|g_{i,j} - g_{i_1,j_1}|$ in the 8-neighborhood of (i,j). These values can be computed at the beginning of the segmentation process using the grey values of the original image (layer $l = 0$). $t_1(l)$ and $t_2(l)$ are thresholds which must be selected experimentally. For $t_1 = 0$ (4) reduces to the constant threshold t_2. In an 8 bit image, where $\text{MIN}\{g_{i,j}\} = 0$ and $\text{MAX}\{g_{i,j}\} = 255$, $t_2 = 5$ is a good choice because grey value differences < 5 cannot be resolved visually. For choosing t_1 the following procedure is under investigation: If one considers a simple image containing a constant grey value plus Gaussian noise, then it seems that the recursion converges for all values of t_1. As in other nonlinear dynamic systems we have different regions of stability. For small values of t_1 ($t_1 < 1$) the algorithm converges to many small segments. At $t_1 \approx 1$ we obtain (together with many very small segments representing noise generated random structures) a big segment which fills out the whole image and represents the background. Here, we also obtain (visually assessed) useful segments for other images whereas for bigger values of t_1 the result becomes too smooth and the useful segments merge to bigger ones which are not in accordance with those which we see. These investigations have not been finished, and we have only preliminary results up to now.

The final PAG (of layer l_{max}) which, in terms of nonlinear systems is an *attractor* for various images, is defined by the node adjacency list $\text{NAL}_g(i,j)$. Besides the segments (as connected components of the PAG) NAL_g also contains edge information.

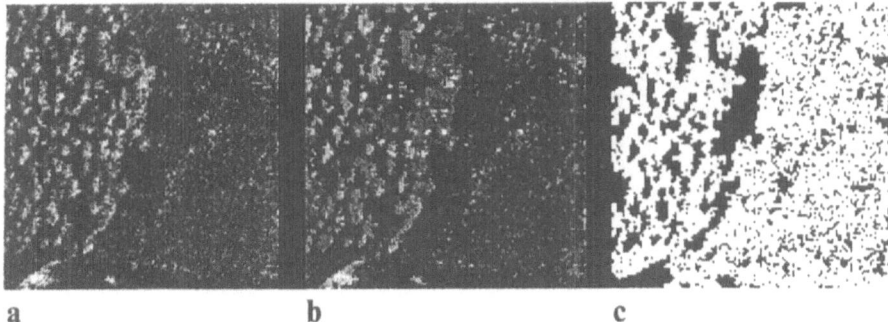

Figure 1. Forest. **a** Original image, **b** segmented image (1393 segments), **c** the biggest segment (10839 pixels)

Figure 1a shows an aircraft image of a forest. The image size is $N = 128$ (16384 pixels). The smoothed grey value image (Fig. 1b) with 1393 segments is an acceptable approximation of the original image. The biggest segment (with 10839 pixels) which represents the background is displayed in Fig. 1c. The many small segments inside the background segment represent tree tops and hopefully can be used as texels for later texture segmentation.

The Mars image of Fig. 2a gives 982 segments (the biggest one with 14125 pixels) for $l = 1$ (Fig. 2b). After 5 recursions we obtain 745 segments (Fig. 2c). The edge image of Fig. 2c (Fig. 2d) shows (1) that the biggest segment represents the whole background, (2) that small details (e.g. craters) are retained, and (3) that there are many isolated noise points (which can be removed in a following cleaning process).

4. Texture Segmentation

In this section we assume that the grey value segmentation process has produced many small segments which can be used as texture elements. Using the NAL, we can compute some features which describe the texels adequately. Simple features which are important for texture discrimination are the size, the orientation, the mean grey value (or color) and the location of a texel. It seems that these features can be understood as textons in the sense of Julesz [2]. Further features which are important are certain descriptors of shape. Because adequate shape description in pre-attentive vision is not yet well understood, we do not consider shape features here. This must be postponed to future studies. The method is demonstrated here for texel location and orientation only. Texel size [9] and grey value can be handled similarly.

4.1. Texel Location

The texel location here is given by the position of the center of gravity which can be computed easily. If we have M texels then the texel positions are given by

H. Jahn

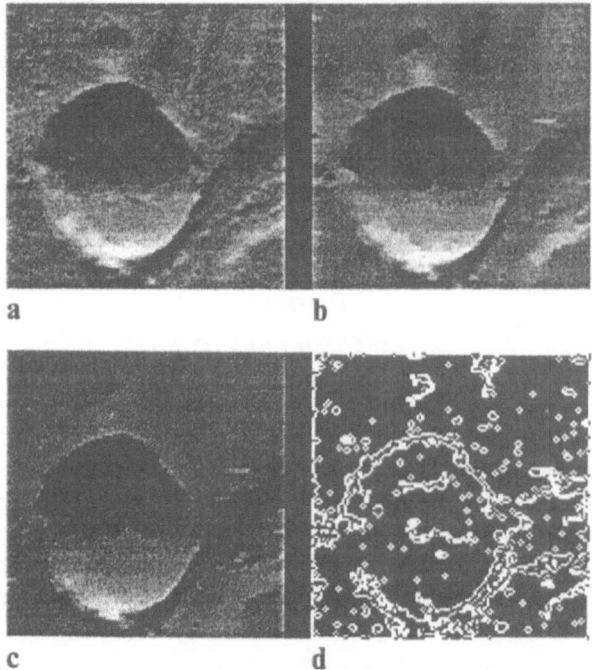

Figure 2. Mars image from Viking spacecraft. **a** Original image, **b** result for $l = 1$, **c** result for $l = 5$, **d** edge image of **c**

$P_k = (x_k, y_k)$ $(k = 0, \ldots, M - 1)$. The points P_k have certain spatial relationships which can be exploited for grouping or clustering these points. In 1985 I have developed a method for dot pattern clustering [11] which can be used here for the segmentation of the texel location set $\{P_k\}$. This cluster algorithm is founded on a proximity criterion and a graph structure fitting the general framework developed in Section 2.

Let us consider three points A, B, C (Fig. 3). In Fig. 3a (left) the distances $d(A, B)$ and $d(B, C)$ are comparable and no separation of the points is perceived. But in Fig. 3b the distance $d(A, B)$ is much greater than $d(B, C)$. Now a grouping is perceived: B and C group together and A is perceived as an isolated point.

Starting from that observation we can formulate the following proximity criterion: Let d_p be the distance from point P to its nearest neighbor. Then two points

Figure 3. Perception of dot patterns

P and P' which are the nodes of a Point Proximity Graph (PPG), a special variant of the FSG, will be connected by an edge if the condition

$$d(P, P') \leq \mu \cdot \text{MIN}\{d_P, d_{P'}\} \qquad (5)$$

is fulfilled. If for μ a value slightly smaller than 2 is chosen then it is guaranteed that in Fig. 3a the nodes A, B and B, C become connected but not A, C. This choice of μ seems also to be in accordance with human perception. (5) is a special version of the feature similarity criterion (1), the feature difference $|f - f'|$ here is replaced by the Euclidean distance, and the feature variation is given by the minimum of nearest neighbor distances. It can be shown that (5) explains the Gestalt phenomenon "grouping by proximity".

In [11] the clustering method roughly outlined here was applied to various dot sets. It showed a good grouping capability which was often in accordance with human perception, but there were also some cases where the method was not able to segregate visual separate clusters. A bridge between a single pair of points can be responsible for that behaviour. Therefore, as for grey value segmentation, a recursive application of (5) together with an adequate smoothing procedure seems to be necessary in order to include regional and global processing.

Figure 4 shows a dot set which (for $l = 1$) is successful separated into two clusters. Figure 4 makes clear that the method can cope with a slight variation of the point distance inside a cluster. Only abrupt distance changes lead to separation. This is in accordance with perception.

Another example (Fig. 5) shows the ability of the method to separate textures with different spatial frequency. Therefore, no Fourier or other transform is needed. Local connections are sufficient. Of course, a few points violating (5) between two regions bridge them (Fig. 5b), but in this case a strong (interrupted) edge between the clusters can be used for object recognition. As for grey value segmentation, a recursive graph structure is able to diminish this problem see Figs.

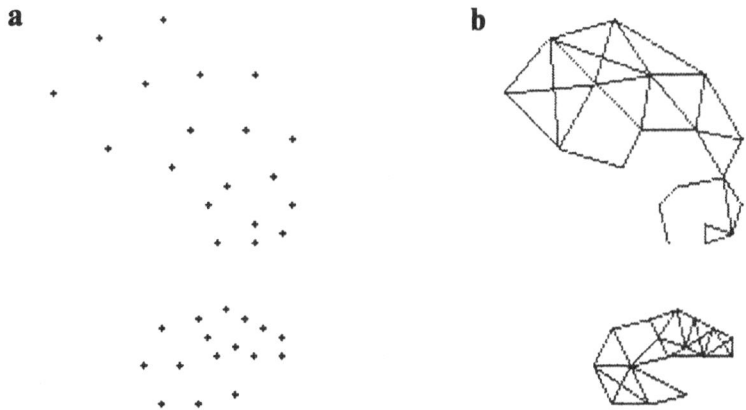

Figure 4. A dot set (**a**) and the Point Proximity Graph (**b**)

H. Jahn

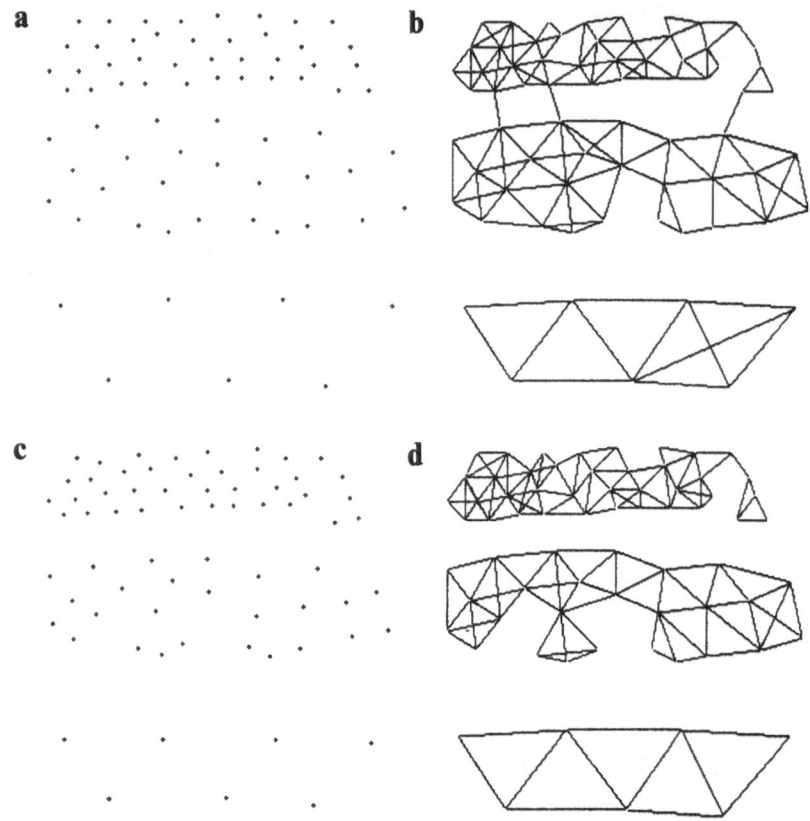

Figure 5. Textures with different spatial frequency (**a**), **b** point Proximity Graph for ($l = 1$ and $\mu = 1.99$), **c** smoothed dot pattern, **d** final result

5c (smoothed dot pattern) and 5d (final segmentation result). This must be investigated further, especially the dot pattern smoothing procedure.

4.2. Texel Orientation

Texture segmentation by orientation of texels was intensively studied by Nothdurft [3]. Important results of these studies are: (1) feature contrast is essential for segregation, target salience increases with feature contrast, (2) continuous variations in texel orientation do not provide segregation, (3) **target salience decreases with increasing background variation.** Using the developed model, it is no problem to explain these observations. To apply the method, a direction ϕ_k is assigned to each texel k with position P_k. ϕ_k is calculated as the angle between the main axis of inertia and the x-axis. Then, for each texel k we consider all texels k' which are located in a certain neighborhood N_k of P_k. For these texels k' the minimal value $\Delta\phi_k$ of the differences $|\phi_k - \phi_{k'}|$ is computed and assigned to texel k.

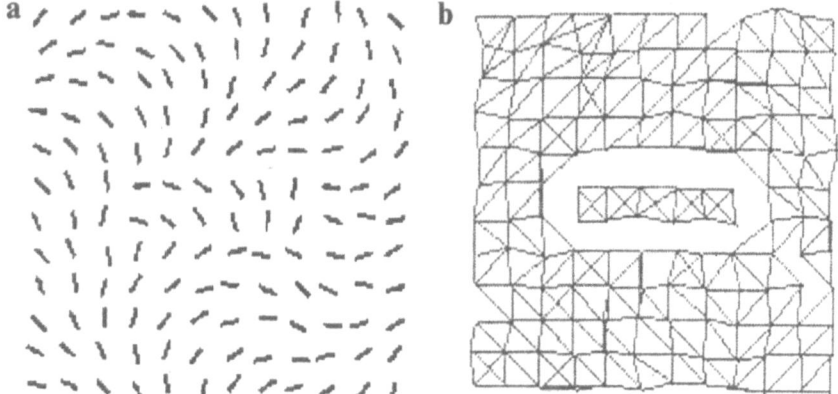

Figure 6. A texture with orientation contrast (a), b orientation similarity graph

Now, let us consider a texel k: For each texel k' being located in a certain neighborhood of P_k (I used a neighborhood with $1,99 d_{P_k}$) the orientation similarity criterion

$$|\phi_k - \phi_{k'}| \leq \text{MAX}\{t_1 \cdot \text{MIN}[\varDelta\phi_k, \varDelta\phi_{k'}]; t_2\} \qquad (6)$$

is applied. If (6) is fulfilled then texel k' is added to the neighbor adjacency list $\text{NAL}_\phi(k)$ which defines the Orientation Similarity Graph (OSG). In (6) there are two thresholds which have been determined experimentally.

The application of the algorithm to the pattern of Fig. 6a (taken from Nothdurft [3]) gives two segments which is in accordance with the human segregation ability. This example also shows that a smooth variation of orientation inside the object and the background does not lead to a partition of background and target, respectively. If the variation of orientation in the background and/or in the target is enhanced then according to (6) the separability of the target is diminished. If the variation is too big then the segments merge. It should be pointed out that the Gestalt law "grouping by similarity" can be explained by the OSG based on (6).

Texture size can be handled similarly. Figure 7 is an example for segmentation by texel size.

5. Conclusions

Ideas for an unified method for grey value and texture segmentation have been developed. It could be shown that with structurally equal criteria (compare (4), (5) and (6)!) graphs can be constructed which allow a successful grouping of pixels and texels. At the moment, the grey value segmentation part of the graph network has reached a state where an application to real world images is possible. Various images have been processed with acceptable results (only two could be shown here

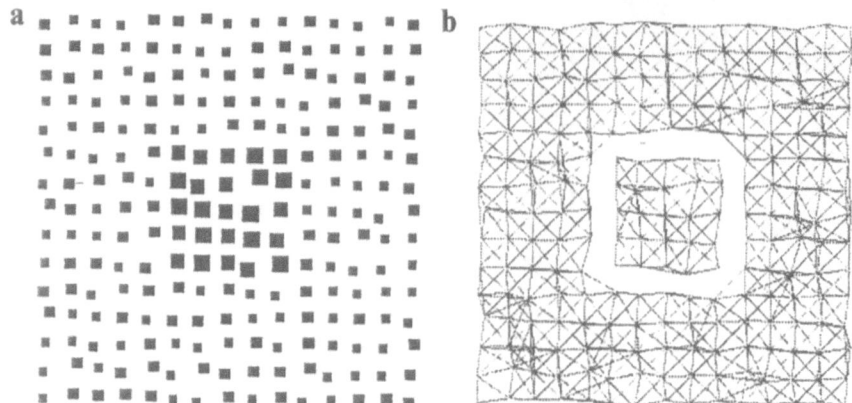

Figure 7. A texture with size contrast (**a**), **b** segmentation result

because of space limitation). Up to now, the texture segmentation part was applied only to simulated images which are used by psychophysicists for investigations of human perception. It was shown that the method can explain many experimental results obtained in this field. To apply the method to real world textures (as shown in Fig. 1) some problems must be solved. First of all, a proper shape description for pre-attentive vision must be developed and a shape similarity criterion has to be formulated.

References

[1] Haralick, R. M., Shapiro, L. G.: Image segmentation techniques. CVGIP *29*, 100–132 (1985).
[2] Julesz, B.: Textons, the elements of texture perception, and their interactions. Nature *290*, 91–97 (1981).
[3] Nothdurft, H. C.: Feature analysis and the role of similarity in preattentive vision. Percept. Psychophys. *52*, 355–375 (1992).
[4] Haralick, R. M.: Statistical and structural approaches to texture. Proc. IEEE *67*, 786–804 (1979).
[5] Jolion, J. M., Montanvert, A.: The adaptive pyramid: a framework for 2D image analysis. CVGIP *55*, 339–348 (1992).
[6] Kropatsch, W. G., Yacoub, S. B.: A revision of pyramid segmentation. Proc. ICPR'96, 477–481 (1996).
[7] Jolion, J. M., Rosenfeld, A.: A pyramid framework for early vision. Dordrecht: Kluwer 1994.
[8] Jahn, H.: Image segmentation with a layered graph network. SPIE Proc. *2662*, "Nonlinear Image Processing VII", 217–228 (1996).
[9] Jahn, H.: A graph structure for image segmentation. SPIE Proceedings *3026*, "Nonlinear Image Processing VIII" (1997).
[10] Pavlidis, T.: Structural pattern recognition. Berlin Heidelberg New York: Springer 1977.
[11] Jahn, H.: Eine Methode zur Clusterbildung in metrischen Räumen. Bild und Ton *39*, 362–370 (1986).

H. Jahn
Deutschs Zentrum
für Luft- und Raumfahrt e.V. (DLR)
Institut für Weltraumsensorik
Rudower Chaussee 5,
D-12489 Berlin
Germany
e-mail: herbert.jahn@dlr.de

Computing Suppl 12, 83–92 (1998)

Discrete Maps: a Framework for Region Segmentation Algorithms

L. Brun, J.-P. Domenger, and **J.-P. Braquelaire,** Talence

Abstract

In this paper, we present different recent segmentation works based on discrete maps. Discrete maps provide an efficient framework for region based segmentation methods. A discrete map is a mixed model combining an encoding of the discrete boundaries of the image regions with topological graphs which represent the topology of the image.

AMS Subject Classifications: 68R05, 68R10, 68U05, 68U10.

Key words: Topological graph, discrete map, segmentation.

1. Introduction

Region-based segmentation algorithms are based on a partition of an image into nonempty connected sets of pixels, called *regions*. The segmentation processing consists in modifying the partition by *splitting* a region into sub-regions or by *merging* two adjacent regions. Region selection is performed using an evaluation criterion computed on one region or on two adjacent regions. Among region-based segmentation algorithms, we distinguish three main approaches:

- The *top-down* approach [15, 7] starts with an under-segmentation of the image and iterates split operations;
- The *bottom-up* approach [3, 10] starts with an over-segmentation of the image and iterates merge operations;
- The *mixed* approach [11, 8, 16] modifies the partition by combining split and merge operations.

The data structures commonly used in the top-down approach are hierarchical structures like *quadtrees* [9, 17] or *pyramids* [4, 16]. The splitting is efficiently performed by a refinement of the subdivision, but the computation of the adjacency of regions involves complex and costly processing. Another drawback implied by a regular subdivision is the "square aspect" of the final segmented image. The bottom-up approach is frequently implemented by *an array of labels* [14] combined with a *region adjacency graph* (RAG) [8, 3]. The mixed approach is usually implemented by a combination of *regular pyramids* and RAG. In this case, the region merging must preserve the hierarchical structure and is called *restricted merge*. Due to the incompatibility of data structures used by the split and the merge algorithms, the mixed approach is mainly divided in two parts: the first part alternates splits and restricted merges; the second part generates a RAG and allows merging of any adjacent regions.

Another region-based approach is proposed by Jolion [12]. This approach is based on adaptive pyramids. During a first stage (bottom-up) an adaptive pyramid is built. The roots of the pyramid define the regions of the image. A second stage (top-down), allows the construction of a graph describing the components.

In this paper, we sum up different recent segmentation works [2, 5, 6] based on a data structure, called *discrete maps* [2], that combines an explicit encoding of the geometry of region boundaries with a topological description based on *topological graphs*. This data structure is well-adapted to mixed approach because the split and merge algorithms use the same data structure and these algorithms can be alternated without overhead. Moreover, this structure allows us to consider regions as 4-connected set of pixels having free geometry.

Section 2 briefly sums up the model of discrete maps. Section 3 presents the ability of this model to be used in a mixed approach. Section 4 is devoted to the presentation of an example of segmentation using the model of discrete maps.

2. Representation of an Image with Discrete Maps

The topology of the subdivision of the Euclidean plane into connected components can be represented by a *topological graph* [18]. The aim of *discrete maps* is to associate a topological graph to a segmented image so that each face of this graph is homeomorphic to one region of the segmented image. To obtain this result, it is necessary to embed the segmented image into a digital topological plane having correct topological properties. The following properties:

1. The boundary of a region r, denoted by $\partial(r)$, is made of discrete Jordan curves.
2. If r and r' are two 4-adjacent regions of the image, then $\partial(r) \cap \partial(r')$ is a set of Jordan arcs. Each arc contains at least three points.

have been established by Braquelaire et al. [2] by embedding the image in the plane $(\frac{1}{2}\mathbb{Z})^2$ provided with Khalimsky topology [13].

These both properties allow us to define the topological graph of a segmented image, if this image is subdivided into simply connected regions. The *faces* of this graph correspond to regions of the images; the edges correspond to Jordan arcs (called *segments*) shared by two 4-adjacent regions; the *vertices* are either points belonging at least to the boundaries of three regions, or arbitrary points selected on a closed Jordan arc included in the boundary of two regions. The geometrical element of a vertex is called a *node*.

A *discrete map* is defined as the drawing of a topological graph in the Khalimsky plane. If the image has n holed regions, its boundary set is the drawing of $n + 1$

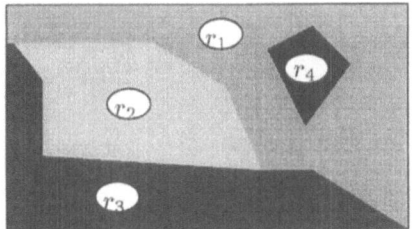

(a) Segmented Image

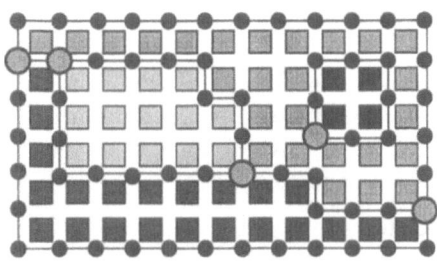

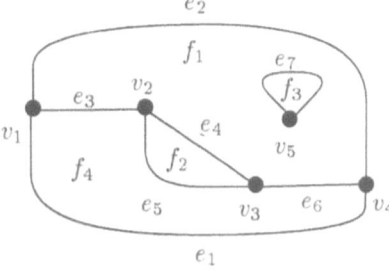

(b) geometry of the discrete maps   (c) Topological graphs of the discrete maps

Figure 1. a shows the segmented image composed by four regions respectively labeled by r_1, r_2, r_3, r_4. **b** represent the boundary set of regions. The big circles are the nodes and the black circles are boundary points of segments. **c** describes the topological maps of the image, there exists two connected components in the graph because one region is holed

discrete maps. We call the *infinite face* of a discrete map, the face representing the exterior of this discrete map. Hence, an image can be represented by a set of discrete maps augmented with an inclusion relation linking each infinite face with the finite face that contains it.

Note that is not necessary to encode the geometry of the discrete maps in the plane $(\frac{1}{2}\mathbf{Z})^2$. In fact, it is sufficient to encode the boundary of regions in the plane corresponding to the closed points of the Khalimsky topology. This plane, called the *boundary plane*, is generated from $\mathbf{Z}^2$ by a a translation of a vector $(-1/2, -1/2)$. The geometry of a discrete map encoding an image of size $h \times w$, is represented by an array $\mathscr{B}$ having a size $(h+1) \times (w+1)$. Let p be a boundary point; the value $\mathscr{B}[p]$ encodes the links which connect p to its neighboring boundary points. Thus, four bits are sufficient to encode an entry of $\mathscr{B}$ [2]. Figure 1 presents the discrete maps associated with an image composed with four regions labeled by $\{r_1, r_2, r_3, r_4\}$.

A topological graph can be efficiently encoded using a combinatorial map. A combinatorial map is a tuple $(\mathscr{D}, \alpha, \sigma)$ such that $\mathscr{D}$ is a set of elements called darts, α and σ are two permutations defined on $\mathscr{D}$. Each dart represents an oriented edge of the topological map. An edge of the topological graph corresponds to an orbit of α, a vertex corresponds to an orbit of σ, an orbit of permutation $\varphi = \sigma \circ \alpha$ encodes a face. The permutation α associates to each dart the dart representing

L. Brun et al.

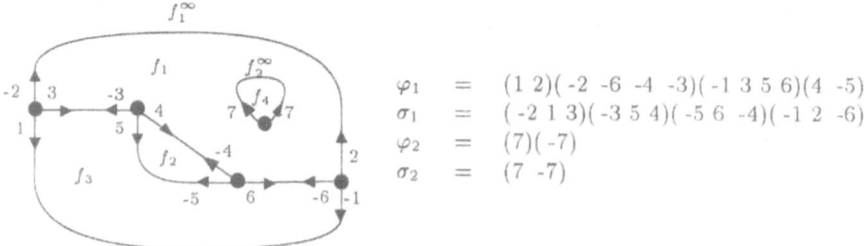

$$\begin{aligned}
\varphi_1 &= (1\ 2)(\text{-}2\ \text{-}6\ \text{-}4\ \text{-}3)(\text{-}1\ 3\ 5\ 6)(4\ \text{-}5)\\
\sigma_1 &= (\text{-}2\ 1\ 3)(\text{-}3\ 5\ 4)(\text{-}5\ 6\ \text{-}4)(\text{-}1\ 2\ \text{-}6)\\
\varphi_2 &= (7)(\text{-}7)\\
\sigma_2 &= (7\ \text{-}7)
\end{aligned}$$

Figure 2. The two combinatorial maps on the image displayed in Fig. 1. For instance, we have $e_1 = \alpha_1^*(1)$, $v_2 = \sigma_1^*(-3)$, the face $f_1 = \varphi^*(-2) = (-2-6-4-3)$

the same edge with an opposite orientation. Thus the permutation α is an involution and if the darts are encoded by positive and negative integers, the permutation α can be implicitly encoded by $\alpha(d) = -d$. Let d be a dart and π a permutation, the orbit of π generated from d is denoted by $\pi^*(d)$.

A labeling function λ defined on $\mathscr{D}$ associates a same label to each dart of one face. Thus, each face is identified by a *label* and conversely the function λ^{-1} associates the label of a face with a canonical dart.

The inclusion relation between discrete maps is encoded by the function **Parent** and **Children** defined on the set of face labels. Let f be a finite face and $\{f_1^\infty, \ldots, f_n^\infty\}$ be the set of infinite faces included in f; then we have

$$\textbf{Children}(f) = \{f_1^\infty, \ldots, f_n^\infty\} \ \text{and} \ \forall f_i^\infty, \textbf{Parent}(f_i^\infty) = f$$

The combinatorial maps representing the topology of the image showed in Fig. 1 are displayed in Fig. 2. The two connected components of the graph are respectively associated with the combinatorial maps (σ_1, α_1) and (σ_2, α_2). The infinite face of the combinatorial map (σ_2, α_2) is $\varphi_2^*(-7) = (-7)$, so we have $\lambda(7) = f_2^\infty$ and **Parent**$(f_2^\infty) = \lambda(-2) = f_1$.

Using the array $\mathscr{B}$, an oriented segment S_d corresponding to a dart d may be traversed starting from its node n if we know its first move m. The association between the topology and the geometry of the discrete maps is given by the functions Σ and Σ^{-1} such that $\Sigma[(n, m)] = d$ and $\Sigma^{-1}[d] = (n, m)$. Let r be a simply connected region labeled by f_r, if $\varphi^* = (\lambda^{-1}(f_r)) = (d_1, \ldots, d_n)$ then

$$\partial(r) = S_{d_1} \oplus \cdots \oplus S_{d_n}$$

where $\oplus$ symbolizes the concatenation of oriented segments. In the case where the region is not simply connected, the other curves of the boundary are associated in the same way with the cycle of infinite faces $\varphi^*(\lambda^{-1}(f_i^\infty))$, with $f_i^\infty \in \textbf{Children}(f_r)$. Since each closed boundary component validates the Jordan theorem, the region r can be traversed using the scan-line or seed-fill algorithms defined in [2].

The model of discrete maps allows us to compute, on a region, a complex homogeneity criteria. We denote by $\mathscr{I}(r)$, the information attached to a region r which allows us to compute an homogeneity criterion. For instance, the homogeneity criterion defined by Beverige [1] requires, for a region r belonging to a grayscale image,

- μ, the mean gray value of r;
- σ, the standard deviation;
- $|r|$, the number of pixels of r;
- $\partial(r)$, the boundary of r;
- $|\partial(r)|$, the perimeter of r;
- $|\partial(r) \cap \partial(r_i)|$, the length of the boundary shared by r and one of its adjacent region r_i.

The information $\mathcal{I}(r)$ can be subdivided in two parts;

1. Region part: $(\mu, \sigma, |r|)$ associated with the label f_r;
2. Boundary part: $(\partial(r))$ distributed on each segment S_{d_i} such that each di belongs either to $\varphi^*(\lambda^{-1}(f_r))$ or to $\varphi^*(\lambda^{-1}(f_i^\infty))$ with $f_i^\infty \in$ ***Children***(f_r).

3. Region-Based Algorithms with Discrete Maps

Region-based algorithms are implemented with two operations, region splitting and region merging. The selection of the operation depends on two homogeneity criteria C_s and C_m. The criterion $C_s(r)$ determines the splitting of a region r; r should be split if $C_s(r) = false$. The criterion $C_m(r, r')$ determines if two adjacent region r and r' should be merged; r and r' are merged if $C_m(r, r') = false$. We suppose that the computation of these both criteria depends on the information $\mathcal{I}$ associated with regions. An image is wholly segmented when both of the following conditions are validated:

1. For all regions r_i, $C_s(r_i) = true$
2. For all regions r_i, $C_m(r_i, r_j) = true$ with r_j an adjacent region of r_i.

A split and merge algorithm consists in the selection of the most heterogeneous region. The selected region is split into $\{r_1, \ldots, r_k\}$. Then, for each region r_i we merge r_i with one of its adjacent region r_j if $C_m(r_i, r_j) = false$. This algorithm is applied recursively until both of the previous conditions are validated.

The main requirements for the split algorithm are:

1. Select a region r labeled by f_r such that $C_s(r) = false$;
2. Traverse r in order to produce the list $\{S_1, \ldots, S_n\}$ of splitting segments;
3. Insert each segment S_i in the data structure;
4. Compute the list of new regions $r_1, \ldots, r_k$;
5. Update the inclusion relation between discrete maps;
6. Evaluate $\mathcal{I}(r_i)$ for each new region r_i.

As mentioned in Section 2, the darts defining the boundary $\partial(r)$ are computed by traversing the orbits of $\varphi^*(\lambda^{-1}(f_r))$ and $\varphi^*(\lambda^{-1}(f_i^\infty))$ with $f_i^\infty \in$ ***Children***(f_r). The insertion of segment S in the geometry data structure is done in a time complexity equal to $|S| + |S_{mean}|$, where $|S_{mean}|$ is the mean length of segments (see [5] for further information). The insertion of an edge in the combinatorial map is direct. The updating of the inclusion relation between discrete maps, requires for each infinite face $f_i^\infty \in$ **Children**(f_r) to update the function **Parent**. The function **Parent**(f_i^∞) is updated by the localization of the region containing a point belonging to the boundary of f_i^∞. The time complexity of the point localization is bounded by the diameter of the region r. The evaluation of $\mathscr{I}(r_i)$ may imply a traversal of the region r_i. Note that the information evaluated on the boundary can be computed during insertion of the segment. If we except this last requirement, the time complexity of the split is lower than $2 \times |r|$.

The main requirements for the merge algorithm are

1. Select a region r labeled by f_r;
2. Compute the adjacency-region set $\mathscr{V}(f_r)$ of the region r;
3. Select a region $r_i \in \mathscr{V}(f_r)$ such that $C_m(r, r_i) = false$;
4. Suppress the edges belonging to $\partial(r) \cap \partial(r_i)$
5. Update $\mathscr{I}(r) = \mathscr{I}(r \cup r_i)$.

The adjacency-region set is computed on the data structure by the adjacency relation between faces expressed by the following definition

Definition 1. *Two finite faces f and f' are adjacent if they verify one of the following properties:*

1. $\exists d \in \varphi^*(\lambda^{-1}(f))$ *such that* $-d \in \varphi^*(\lambda^{-1}(f'))$
2. $\exists d \in \varphi^*(\lambda^{-1}(f))$ *and* $\exists! f^\infty \in$ ***Children***(f') *such that* $-d \in \varphi^*(\lambda^{-1}(f^\infty))$
3. $\exists d \in \varphi^*(\lambda^{-1}(f'))$ *and* $\exists! f^\infty \in$ ***Children***(f) *such that* $-d \in \varphi^*(\lambda^{-1}(f^\infty))$

The neighborhood of one face defined by definition 1 can be computed thanks to the algorithm displayed in Fig. 3. An example of neighboorood is displayed in the same figure.

The time complexity of the computation of $\mathscr{V}(f_r)$ is linear in $\sum |\varphi^*(d_i)|$ where $d_0 = \lambda^{-1}(f_r)$ and $d_i = \lambda^{-1}(f_i^\infty)$. Note that, with the same algorithm, we identify for each face $f_{r_i} \in \mathscr{V}(f_r)$ the edges belonging to $\partial(r) \cap \partial(r_i)$. The removal of an edge in the topological graph is direct, and the removal of a segment S is in $|S|$ because S must be traversed in order to be removed in the array $\mathscr{B}$ [5]. The updating of the inclusion relation is direct. If the information attached to two adjacent regions r and r' verifies $\mathscr{I}(r \cup r') = \mathscr{I}(r) \cup \mathscr{I}(r')$, then the fourth requirement of the merge can be done immediately.

4. A Split and Merge Algorithm Using Discrete Maps

The homogeneity criterion, used by the split and merge algorithm presented in [6], is based on the squared error defined by:

$$SE(FS_r) = \sum_{x \in FS_r} g(x) \|x - \mu(r)\|^2 \tag{1}$$

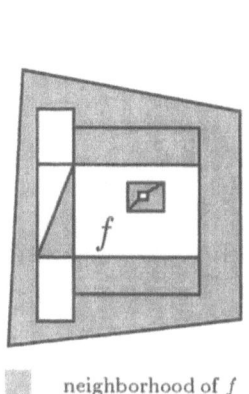

```
compute_neighbor(face f,list of faces L)
{
        dart d,d_first
        face f'

        d_first = d = λ⁻¹(f)
        do {
                if(Parent(λ(-d)) == ∅)
                        f' = λ(-d)
                else
                        f' = Parent(λ(-d))
                L = L ∪ {f'}
                d = φ(d)
        } while(d! = d_first)
        if(Parent(f) == ∅)
                for all f' in Children(f)
                        compute_neighbor(f',L)

}
```

neighborhood of f

Figure 3. The neighborhood of the face f displayed on the left part of this figure is defined by Definition 1. This neighborhood may be computed thanks to the algorithm compute neighbor, displayed on the right part of this figure. This algorithm store in a list L the neighborhood of a face f. Since a finite face has no parent face we have **Parent**$(f) = \emptyset$

where FS_r denotes the feature space associated to region r, this feature space may be a set of values between $\{0, \ldots, 255\}$ for grayscale images or a three dimensional set for color images. The value $g(x)$ is the number of pixels with color x in region r and $\mu(r)$ is the mean of the set FS_r. Now, we will denote $SE(r)$ for $SE(FS_r)$. The criterion C_r is defined by

$$C_s(r) = false \Leftrightarrow SE(r) > \alpha SE(I)$$

where α is a parameter given by the user, and $SE(I)$ is the square error of the initial image. At each step this algorithm split the region r_{max} having the greatest squared error. The splitting of r_{max} generates a set of sub-regions $\{r_1, \ldots, r_n\}$. Then, we apply a merger on the regions $\{r_1, \ldots, r_n\}$ using the criterion C_m defined by:

$$C_m(r_i, r) = false \Leftrightarrow \begin{cases} SE(r \cup r_i) = Min_{j \in \{1,\ldots,n\}, r \in V(r_j)} SE(r_j \cup r) \\ \text{and} \\ SE(r \cup r_i) > \beta SE(I) \end{cases}$$

Where $\beta < \alpha$ may be defined by user or deduced from α. Our Experiments showed us that $\beta = 1/120\alpha$ provides good results for a large range of images.

Note that this merge algorithm may merge one region r_i with a region which does not belong to $\{r_1, \ldots, r_n\}$. The resulting region will be added to the set

```
split_and_merge (region r_max)
{
        list of regions L', L = ∅
        while(SE(r_max) > αSE(I))
        {
            L' = split(r_max)
            L = L ∪ L'
            merge(L', L)
            r_max = ArgMax_{r ∈ L} SE(r)
        }
}
```

Figure 4. This Algorithm resumes the main steps of our split and merge algorithm. The list L corresponds to the set of regions created by the split and merge algorithm, while the list L' corresponds to the list of regions created by the last splitting step

(a) Initial image

(b) First splitting step
2342 Nodes, 3662 Segments
2935 Finite faces, 1064 infinite faces

(c) First merging step
36 Nodes, 54 Segments
19 Finite faces

(d) Final image obtained after twenty iterations
515 Nodes, 778 Segments
288 Finite faces, 24 Infinite faces

Figure 5. This figure shows different steps of the segmentation process and indicates the main parameters of the discrete map model

$\{r_1, \ldots, r_n\}$. In order to restrict the set of regions considered by the merge algorithm, we use an array of marks indicating if a region can be merged or not.

The algorithm displayed in Fig. 4 restricts the merger to the set of regions created by the split and merge algorithm. Therefore, a region generated from a split step may be reconsidered in the following merge steps. The results of this algorithm on the well known test image lenna are displayed in Fig. 5.

5. Conclusion

The model of discrete maps provides a framework to implement any region-based segmentation algorithms. The homogeneity criterion may be computed from information defined on regions or region boundaries. Moreover, this model can be used in an interactive segmentation processing. Thus, the user may reduce the segmentation to an image part having free geometry and not necessary connected. He may also modify the result of a segmentation by incremental addition or removal of segments.

References

[1] Beveridge, J. R., Griffith, J., Kohler, R., Hanson, A. R., Riseman, E. M.: Segmenting images using localized histograms and region merging. Int. J. Comput Vision 2, 311–347 (1989).
[2] Braquelaire, J. P., Domenger, J. P.: Representation of region segmented images with discrete maps. Technical report, RR-112797 LaBRI, available on http://www.labri.u-bordeaux.fr/LaBRI/Publications/index.html, June 1996.
[3] Brice, R., Fennema, C. L.: Scene analysis using regions. *Art. Intell. 1*, 205–226 (1970).
[4] Browning, J. D.: Segmentation of pictures into regions with tile-by-tile method. *Pattern Rec. 15*, 1–10 (1982).
[5] Brun, L., Domenger, J. P.: Incremental modifications on segmented image defined by discrete maps. Technical report, RR-112696 LaBRI, available on http://www.labri.u-bordeaux.fr/LaBRI/Publications/index.html, may 1996.
[6] Brun, L., Domenger, J. Ph.: A new split and merge algorithm with topological maps and interpixel boundaries. In: Proc. WSCG '97 1, 21–30 (1997).
[7] Celenk, M.: A color clustering technique for image segmentation. Comput. Vision Gaphics Image Proc. 52, 145–170 (1990).
[8] Cheevasuvut, F., Maitre, H., Vidal-Madjar, D.: A robust method for picture segmentation based on a split and merge procedure. Comput. Vision Graphics Image Proc. *34*, 268–281 (1986).
[9] Dyer, R. C., Rosenfeld, A., Hanan, S.: Region representation: Boundary codes from quadtrees. ACM: Graphics Image Proc. *23*, 171–179 (1980).
[10] Fiorio, C., Gustedt, J.: Two linear time union-find strategies for image processing. Technical Report 375/1994, Technische Universitat Berlin, 1994.
[11] Horowitz, S. L., Pavlidis, T.: Picture segmentation by a tree traversal algorithm. J. Assoc. Comput. Mach. 23, 368–388 (1976).
[12] Jolion, J. M., Montanvert, A.: The adaptative pyramid: A framework for 2d image analysis. Comput. Vision Graphics Image Proc. *55*, 339–348 (1992).
[13] Khalimsky, E., Kopperman, R., Meyer, P. R.: Boundaries in digital planes. J. Appl. Math. Stoch. Anal. *3*, 27–55 (1990).
[14] Nicol, C. J.: A systolic approach for real time connected component labeling. Comput. Vision Image Underst. *61*, 17–31 (1995).
[15] Ohlander, R., Price, K., Reddy, D. R.: Picture segmentation using a recursive region splitting method. Comput. Graphics Image Proc. *8*, 313–333 (1978).

[16] Pietikainen, M., Rosenfeld, A., Walter, I.: Split and link algorithms for image segmentation. Pattern Rec. *15*, 287–298 (1982).
[17] Samet, H.: Region representation: Quadtrees from boundary codes. ACM Graph Image Proc. 23, 163–170 (1980).
[18] Tutte, W. T.: A census of planar maps. Can. J. Math. *15*, 249–271 (1963).

J. Brun
J.-P. Domenger
J. P. Braquelaire
Laboratoire Bordelais de Recherche
en Informatique
Université Bordeaux I
351, cours de la Libération
33405 Talence, France
e-mails: {brun|domenger|braquelaire}@labri.u-bordeaux.fr

Computing Suppl 12, 93–100 (1998)

Image Sequence Segmentation by a Single Evolutionary Graph Pyramid

P. Bertolino and **S. Ribas**, Grenoble

Abstract

In the presented method, an irregular pyramid is used to segment the successive frames of an image sequence: a pyramid is built with the first image of the sequence, and then it is updated from image to image, using a split-and-merge process that takes into account the changes occurred between two successive frames. Thus, the same pyramid structure is used along the sequence, speeding up the process. Stability criteria allow to have the required compromise between speed and quality, i.e. to look at the image evolution at a particular resolution. Thanks to the graph representation, objects obtained in a given image can be tracked along the rest of the sequence.

Key words: Image segmentation, multiresolution, object tracking, graph pyramid.

1. Introduction

Second generation image coding techniques [4, 6] allow to get rid of block effects, since they are based on objects instead of blocks. They also represent a compromise between spatial or spectral approaches and model-based techniques [5]. The irregular pyramid [3, 8] can be used to segment, or more generally to code an image, with the possibility of choosing a particular resolution at which the representation has to be done [1].

With this method, segmenting or coding an image sequence could simply be done by building one pyramid per image. This approach would not be efficient for the following reasons:

- The processing time would be the same for each image of the sequence even if the content of several successive images is the same or very similar;
- No *a priori* relationship could be established between the representation of the image I_t and the representation of the next image I_{t+1};
- If any relationship might be established between I_t and I_{t+1}, both representations would have to stay in memory and time-consuming graph matching would be needed.

We propose a process in which the pyramid structure is constructed once, for the

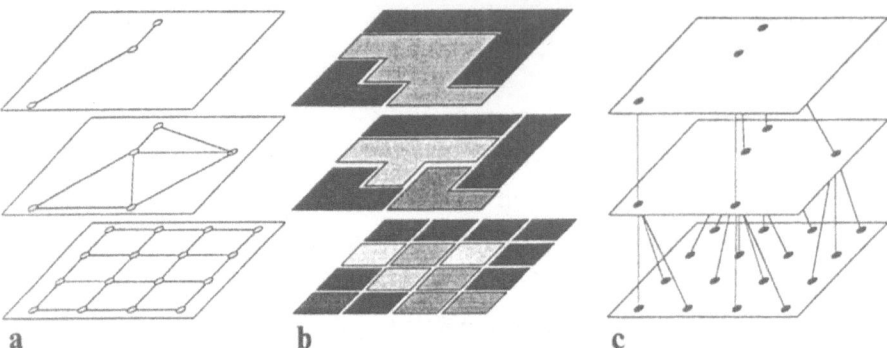

Figure 1. The structures defining the irregular pyramid, with a 4 × 4 image sample. **a** The graphs pyramid, **b** the stack of the receptive fields, **c** the tree structure

first image of the sequence. Then, this hierarchical representation is modified (i.e. updated) in order to take into account the content of the second image, and so on, from an image to the next one. The pyramid updating process is top-down. It is done from the low resolution (apex of the pyramid) to the high resolution (base of the pyramid, i.e. original image).

2. Processing of the First Frame

Processing the first frame is the initializing phase of the method, as it is done in [2]: optimization requires that the pyramid structure must be built once. So, this is done using the first image of the sequence using the following classical steps (Fig. 1):

An adjacency graph (level 0) is derived from the 4 or 8 connexity using the pixels of the original image (Fig. 1a). Then the following steps correspond to the construction of a new level[1]:

Similarity graph From the adjacency graph, a similarity graph is built, taking into account the similarity between each pair of adjacent vertices: according to a similarity criterion (which can be evaluated globally or locally), vertices are linked or not within this graph.

Decimation A local decimation process performed on the similarity graph permits to locally chose the vertex that will represent a part of its neighbourhood on the next level. These vertices (survivors) will be the regions of the next level.

Grouping Non-surviving vertices have to choose in their neighbourhood the most similar surviving vertex to be linked to, in order to form bigger regions in the next level. These relations are inter-levels links or sons-to-father links (Fig. 1c).

---

[1] On the first level, each vertex is a pixel. On the next levels, each vertex is a set of connected pixels (i.e. a region).

Attributes updating Now, surviving vertices can compute their new attributes according to the attributes of their sons: mean grey value, standard deviation of grey levels, area. The connected component attached to a given vertex is its receptive field (Fig. 1b).

Adjacency graph Adjacencies are defined to fit the neighbourhood of the new graph. These edges take into account both the adjacencies of the previous level and the merging among vertices.

This new level is now ready to be processed on its turn to provide the next level and so on, until the apex of the pyramid is reached (i.e. when two successive levels have the same number of regions). For the continuation of the method, every obtained level of this hierarchy is stored in memory.

3. Processing of the Next Frames

Once the pyramid of the first image is built, it must be modified to fit the content of the next image. The smallest possible number of modifications have to be made so that the smallest part of the tree structure has to be visited and updated. In order to take advantage of the hierarchical structure, this fitting is first tested at low resolution, that is on the smaller graph. If locally it cannot be done, it is tried at a higher and higher resolution.

3.1. Splitting Phase

During the bottom-up construction of the pyramid of the first image, links between sons and father are stored so that it can be easy to go down and up through the tree structure thus formed.

The splitting process starts from the apex of the current pyramid (Fig. 2c). Image I_t partition obtained at the apex is applied (traced) to the content of image I_{t+1}. Then, two cases are possible:

1. If the pixels of one region of I_{t+1} look like the ones of the same region of I_t (according to given criteria), the region is said to be *stable* it is not split into sub-regions of the lower level. The corresponding sub-tree and its receptive field will not be checked.
   The criteria are: a mean gray difference threshold, a standard deviation difference threshold, and a threshold area: underneath this threshold, a region is not split anymore, whatever its attributes.
2. If the region is not stable, the same splitting process is iterated on each of the son regions. The regions which are not stable but which cannot split (either they contain just one pixel or their size is lower than the threshold size) are called *modified* regions.

Notice that if two successive images are the same, no split is performed and the new segmentation is directly obtained without no change within the pyramid. In

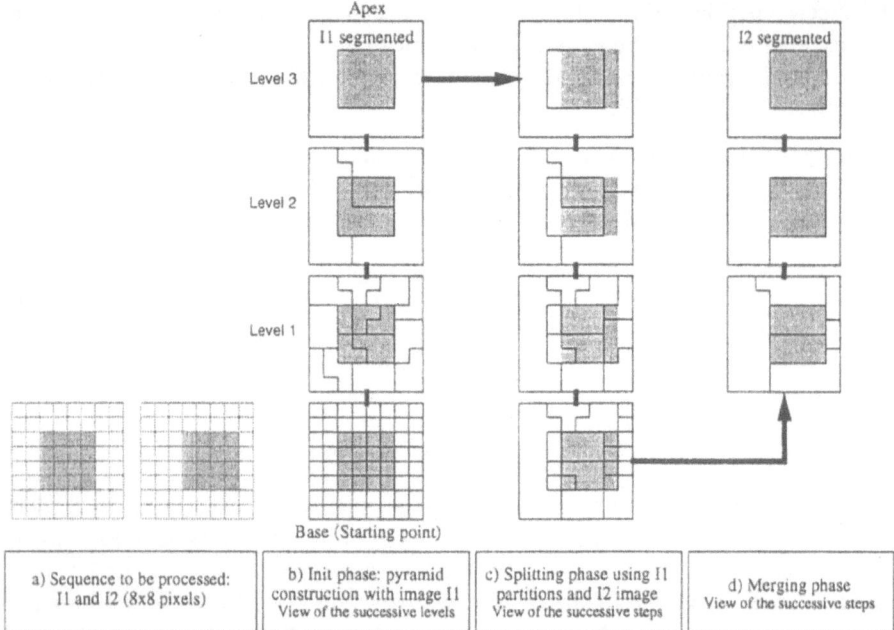

Figure 2. Principle of the method on an example with a two image sequence

the worst case (from a black to a white image for instance), the recursive splits lead to the base of the pyramid without finding any stability.

In the case a modified region does not belong to the base of the pyramid (i.e. it has one or several sons), its attributes will not fit the content of its sons. Nevertheless, this artefact does not affect the results.

3.2. Integration Phase

Integration is the process in which the vertices of the pyramid representing frame I_t are updated with data coming from I_{t+1}. This phase is theoretically performed after the splitting phase. Every modified region (whatever the level it belongs to) is processed. It is updated with attributes corresponding to the pixels of the next image.

Since the splitting phase does not necessary proceed until the bottom of the pyramid, each modified region takes the attributes of the corresponding receptive field of the next image (mean grey level, standard deviation) whatever its value and the similarity within this receptive field.

In order to simplify software implementation, integration of a modified region occurs as soon as the region is known as to be modified, that is during the splitting phase.

3.3. Merging Phase

The upper parts of the tree that have split do not exist anymore. They have to be rebuilt to fit the content of the image as well: starting from the lower level that contains one or more modified regions, the merging process goes up, from level to level.

Only participate to this merging phase the regions which are known as being modified or which belong to the neighbourhood of a modified region. In order to rebuild the whole pyramid, every region that has at least one modified son is on its turn known as modified.

Modified regions are allowed to merge either with a neighbour of the same level or with the father of one of their neighbours (Fig. 2d). Except this merging phase is done using a subset of vertices, it is exactly the same region growing process as stated in Section 2.

It ensures both:

- The refusion of regions that have been disconnected during the splitting process.
- The reconstruction of the upper part of the pyramid taking into account the data coming from the next image.

During this phase, the height of the pyramid may vary, according to the changes and the complexity of the images.

4. Interframes Object Tracking

When the pyramid is created with the first image of the sequence, each of the n regions of the apex (i.e. each vertex) is labeled (from 1 to n). During the splitting phase, each son inherits the label of its father. In the integration phase, every modified region is given the NOLABEL label, meaning that it is not linked to any known object of the apex. During the merging phase, the label of each father must be determined according to the labels of its sons. The following rules are used:

1. If all the sons have label L_i, the father is labeled L_i: the new entity is made of pieces of the same object. Either it is a part of the object, or it is the object itself.
2. If all the sons have either label L_i or label NOLABEL, the father is labeled L_i: the new entity is made of known pieces and ships from an unknown object with similar features (since their fusion is required). For instance, it may result from a slight move of the object.
3. If sons and labeled with different labels (e.g. L_i and L_j), the father is labeled MULTILABEL. This case is the most complicated and is mainly due to problems of segmentation. It is not treated here.

Labeling from sons to fathers is iterated until the apex of the pyramid is reached. Among the vertices of the apex, some have labels (these are regions tracked from

Figure 3. Segmentation of a sequence. 1st column: image sequence. 2nd column: labeled segmentations

the previous image) while others do not have (NOLABEL). These latter regions are given new labels which are not already used for other regions.

The robustness of the method may only be obtained with an adaptive pyramid [3]: the stochastic pyramid [7] prevents fusions to be locally similar when a few details changes between two frames.

5. Results

This method had been tested on 128×128 and 256×256 image sequences. The computational time ratio between the construction of the pyramid of the first image and the updating of the pyramid of each next image shows that the method is well suited for video conference and more generally video sequences (Fig. 3). Updating a pyramid between two frames is 3 to 8 times faster than building the pyramid itself, according to the image and the parameterization. An optimized program should permit to reach ratios higher than 10.

6. Conclusion

We have presented a framework for segmenting image sequences with the powerful irregular pyramid model. This approach may be used in low bitrate coding to compress data. In computer vision, the resolution of the object tracking could be tuned according to the content of the sequence or in order to focus particular objects of the scene. Automated indexing in image data bank could also be investigated.

Though this method requires that objects overlap between successive frames, further work will include motion estimation to overcome this problem. The case of MULTILABEL regions will be taken into account as well.

References

[1] Bertolino, P., Montanvert, A.: Multiresolution segmentation using the irregular pyramid. In: IEEE ICIP, pp. 257–260, Lausanne, Switzerland, September 16–19, 1996.
[2] Gambotto, J. P.: A region-based spatio-temporal segmentation algorithm. In: IEEE proceedings, 11th IAPR, International Conference on Pattern Recognition, Vol. 3 (Likin, B. S., Rosenfeld, A.) pp. 189–192. The Hague, The Netherlands, New York: Academic Press, 1992.
[3] Jolion, J. M., Montanvert, A.: The adapted pyramid: a framework for 2d image analysis. Comput. Vision Graphics Image Proc. 55, 339–348 (1992).
[4] Kunt, M., Ikonomopoulos, A., Kocher, M.: Second-generation image-coding techniques. In: Proc. IEEE 73, pp. 436–440 (1985).
[5] Li, H., Lundmark, A., Forchheimer, R.: Image sequence coding at very low bitrates: A review. IEEE Trans Image Proc. 3, 589–609 (1994).
[6] Marcotegui, B.: Segmentation de séquences d'images en vue du codage. PhD thesis, Ecole Nationale Supérieure des Mines de Paris, 1996.
[7] Meer, P.: Stochastic image pyramids. Comput. Vision Graphics Image Proc. 45, 269–294 (1989).

[8] Montanvert, A., Meer, P., Rosenfeld, A.: Hierarchical image analysis using irregular tessellations. IEEE Trans Pattern Anal. Mach. Intell. *13*, 307–316 (1991).

P. Bertolino
S. Ribas
Laboratorie TIMC-IMAG
Institut Bonniot,
Domaine de la Merci
38706 La Tronche Cedex
France
e-mail: Pascal.Bertolino@imag.fr

Computing Suppl 12, 101 – 110 (1998)

Dual Graph Contraction with LEDA*

W. G. Kropatsch, Vienna, **M. Burge**, Linz, **S. Ben Yacoub**,
and **N. Selmaoui**, Vienna

Abstract

Graphs are useful tools for modeling problems that occur in a variety of fields. In machine vision graph based solutions have been successfully applied to many image processing problems e.g. quad trees for image compression and region adjacency graphs for segmentation. The application of graphs to machine vision problems poses special problems due to the underlying size of the image e.g. a graph representing the base level of a $512x512$ image has over 200.000 nodes. The large size of the graphs make issues of both space and time complexity important when designing algorithms for machine vision problems. We present an implementation under LEDA (Library of Efficient Data structures and Algorithms) of DGC (Dual Graph Contraction) for irregular pyramids. In the first section we present the theory behind DGC, in the second an algorithmic specification is derived, and in the third an implementation under LEDA is given followed by a short conclusion.

Key words: Structural image analysis, graph pyramid, topology preserving.

1. Theory of DGC

The presented approach addresses a representation of pure structure, a hierarchy of plane graphs, with a clear interface, the decimation parameters, to control generation and modification of the structure. Dual graph contraction is the basic process [3] that builds an irregular 'graph' pyramid by successively contracting a dual image graph of one level into the smaller dual image graph of the next level. Dual image graphs are typically defined by the neighborhood relations of image pixels or by the adjacency relations of the region adjacency graph. The above concept has been used for finding the structure of connected components [6]. It also embeds Meer's stochastic pyramid [8], the adaptive pyramid [2], and a further variant of Meer's approach, Mathieu's optimal stochastic pyramid [7] which produced excellent segmentation results by decimating a minimal spanning tree instead of the original graph.

* This work was supported by the Austrian Science Foundation under grant number S7002-MAT.

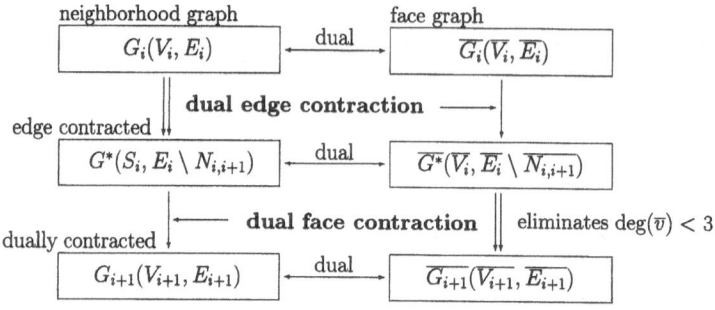

Figure 1. Dual Graph Contraction: $(G_{i+1}, \overline{G_{i+1}}) = C[(G_i, \overline{G_i}), (S_i, N_{i,i+1})]$

Dual graph contraction proceeds in two basic steps (Fig. 1): dual edge contraction and dual face contraction. The base of the pyramid consists of the pair of dual image graphs $(G_0, \overline{G_0})$. Following *decimation parameters* $(S_i, N_{i,i+1})$ determine the structure of an irregular pyramid [3] [Def. 5]: a subset of *surviving vertices* $S_i = V_{i+1} \subset V_i$, and a subset of *primary non-surviving edges*[1] $N_{i,i+1} \subset E_i$. Every non-surviving vertex, $v \in V_i \backslash S_i$, must be connected to one surviving vertex in a unique way. The relation between the two pairs of dual graphs, $(G_i, \overline{G_i})$ and $(G_{i+1}, \overline{G_{i+1}})$, as established by dual graph contraction with decimation parameters $(S_i, N_{i,i+1})$ is expressed by function $C[., .]$:

$$(G_{i+1}, \overline{G_{i+1}}) = C[(G_i, \overline{G_i}), (S_i, N_{i,i+1})] \tag{1}$$

The contraction of a primary non-surviving edge consists in the identification of its endpoints and in the removal of both the contracted edge and its dual edge. Dual face contraction simplifies most of the multiple edges and self-loops, but not those inclosing any surviving parts of the graph (see [3]). One step of dual graph contraction is illustrated in Fig. 2. Note that the contracted graph may contain both a self-loop and multiple edge. They are necessary to preserve the structure defined in thse base graph [3].

To define the parameters that control the process of dual graph contraction we observe that the subgraphs in our example graph (Fig. 2b) form small tree structures $T(s)$ that collapse into surviving vertex s of the contracted graph. $T(s)$ is a *spanning tree* of the connected component of the surviving root vertex, or equivalently, (V, N) is a spanning forest of graph $G(V, E)$ [4].

Definition 1. *A decimation of a graph $G(V, E)$ is specified by a selection of* surviving vertices $S \subset V$ *and a selection of* primary non-surviving edges $N \subset E$ *such that following two conditions are fulfilled:*

---

[1] Secondary non-surviving edges are removed during dual face contraction.

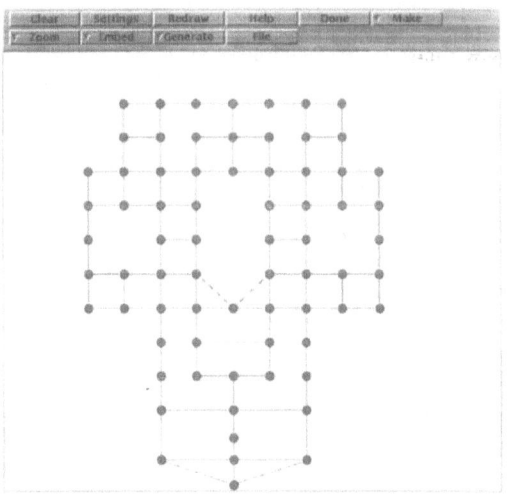

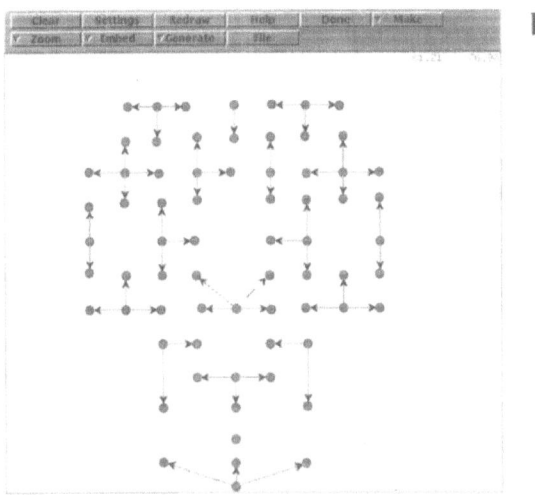

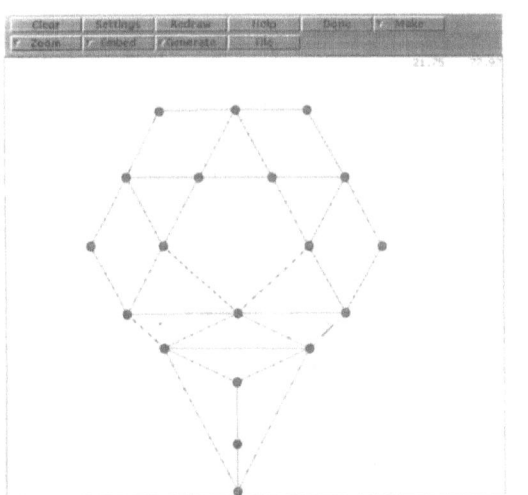

Figure 2. Example of dual graph contraction: $(V_1, E_1) = C[(G_0, \overline{G_0}), (S_0, N_{0,1})]$. **a** (V_0, E_0), **b** $(S_0, N_{0,1})$, **c** (V_1, E_1)

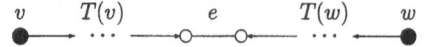

Figure 3. Decomposition of connecting path $CP(v, w)$

1. *Graph (V, N) is a spanning forest of graph $G(V, E)$.*
2. *The surviving vertices $S \subset V$ are the roots of the forest (V, N).*

The trees $T(v)$ of the forest (V, N) with root $v \in V$ are called contraction kernels.

The connectivity structure of the contracted graph is established by paths connecting two surviving vertices:

Definition 2. *Let $G(V, E)$ be a graph with decimation parameters (S, N). A path in $G(V, E)$ is called a* connecting path *between two surviving vertices $v, w \in S$, denoted $CP(v, w)$, if it consists of three subsets of edges E (Fig. 3):*

1. *The first part is a possibly empty branch of contraction kernel $T(v)$.*
2. *The middle part is an edge $e \in E \backslash N$ that bridges the gap between the two contraction kernels $T(v)$ and $T(w)$. We call e the* bridge *of the connecting path $CP(v, w)$.*
3. *The third part is a possibly empty branch of contraction kernel $T(w)$.*

Connecting paths $CP(v, w)$ in $G(V, E)$ are strongly related to the edges in the contracted graph $G'(V', E')$: Two different surviving vertices that are connected by a connecting path in G are connected by an edge in E'. For every edge $e' = (v, w) \in E'$ there exists a connecting path $CP(v, w)$ in G. Dual edge contraction can be implemented by (1) simply renaming all the non-surviving vertices to their surviving parent vertex, (2) deleting all non-surviving edges N and (3) their duals $\bar{N}$.

2. Algorithmic Specification of DGC

In the last section we provided the theory of DGC, from the theory we will now derive an algorithmic specification. The specification is given primarily in the flowchart (Fig. 4) and uses the same notation as in the previous section. The Image to Graph module transforms an image into a representation in which each pixel is a vertex with edges to its four connected neighbors. The dual graph is also created at this time as a separate but linked graph in a way that a syntactic equivalence between the graph and its dual is provided while maintaining the semantic difference. The Dual_Graph_Contraction module with (S, N) decimation (N is oriented as shown in Fig. 2b) parameters illustrates the main algorithm for DGC and calls the following modules: Dual_Edge_Contraction and Dual_Face_Contraction use *Substitute*, which links nodes to the surviving root using an iterative substitution algorithm.

Dual_Edge_Contraction
  Edge_Contract(G,S,N) {
    for_all_edges(e = (v, w), N){
      G.substitute(v,root(v));
      G.substitute(w,v);
      G.del_edge(e);
    }
    for_all_edges(e = (v, w), G){
      G.new_edge(
        e' = (root(v), root(w)));
    }
  }
  Edge_Remove($\bar{G}, \bar{N}$){
    for_all_edges($\bar{e}, \bar{N}$)
$\bar{G}$.del_edge($\bar{e}$);
  }
Dual_Face_Contraction {
F is set of faces to contract
M set of edges in $\bar{E}$ to contract
  Dual_Face_Select($\bar{G}, F, M$){
    for_all_edges($\bar{v}, \bar{G}$){
      if deg($\bar{v}$) ≥ 3 **then** {
        F.append($\bar{v}$);
        M.insert($\bar{G}$.adj_edge(*barv*));
      }
    Edge_Contract($\bar{G}, F, M$);
    Edge_Remove(G, $\bar{M}$);
}

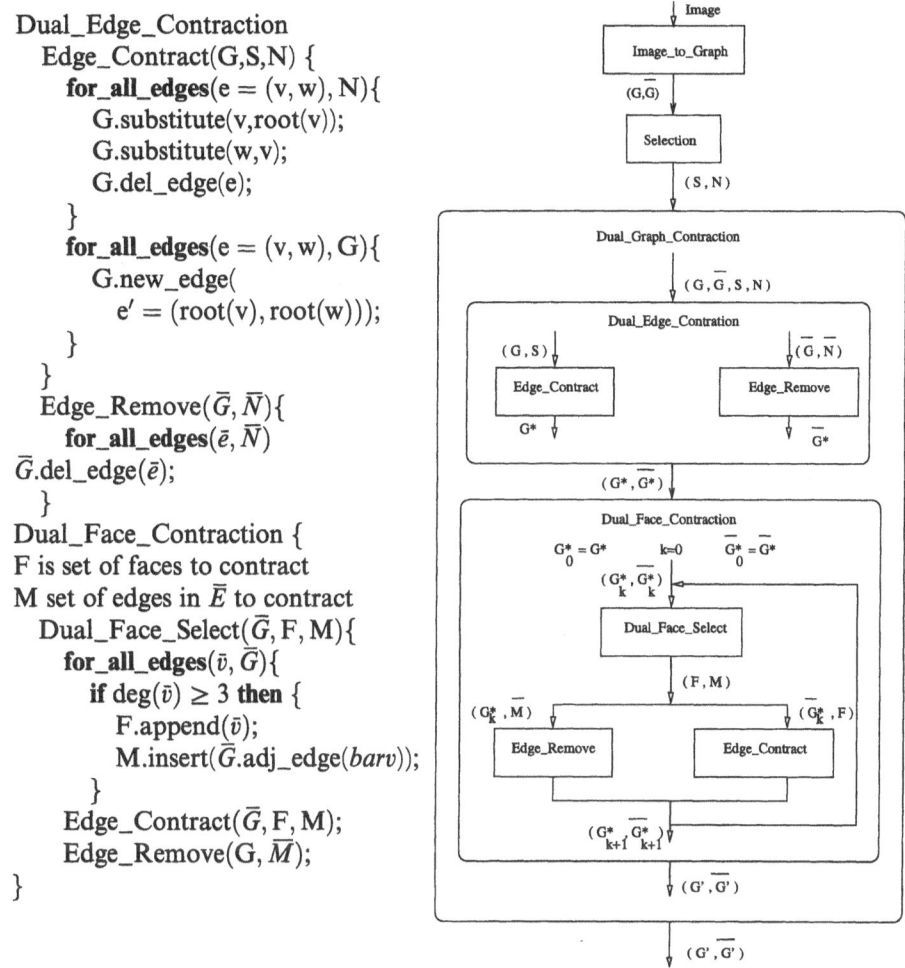

Figure 4. Algorithmic specification of dual graph contraction

3. Implementation of DGC in LEDA

LEDA [9] (Library for Efficient Data types and Algorithms) is a C++ class library implementing many common abstract data types e.g. trees, graphs, and lists. The graph support in LEDA includes iterators like "**forall_nodes** of graph G do", and "**forall_adj_edges** of v do" as well as basic operations like "delete node" and "delete edge". An instance G of the data type *graph* consists of a list V of nodes and a list E of edges, where node and edge are LEDA classes. A pair of nodes $(v, w) \in V \times V$ is associated with every edge $e \in E$, v is called the *source* of e, w the *target* of e, and v and w are the *endpoints* of e. In LEDA a graph is either *directed* or *undirected* where the difference is in the way the edges incident to a node are stored and how the concept of *adjacent* is defined. In directed graphs two lists of edges are associated with every node v: adj_edges $(v) = \{e \in E \mid v =$

source(*e*)}, i.e., the list of edges starting in v, and in_edges $(v) = \{e \in E \mid v = $ *target*(*e*)}, i.e., the list of edges ending in v. The list adj_edges (v) is called the adjacency list of node v and the edges in adj_edges (v) are called the edges *adjacent* to node v. For directed graphs we often use *out_edges*(*v*) as a synonym for *adj_edges*(*v*).

LEDA takes advantage of $C++$ templates to allow for truly generic algorithms. Templated graphs are similar to attributed graphs, in an attributed graph, for every vertex v and every edge e of $G(V, E)$, there exists a function $f(v)$ which returns the attribute of vertex v and $g(e)$ which returns the attribute of edge e, in $C++$ notation that would be GRAPH G<V,E>, with G[v] returning the attribute of vertex v and G[e] returning that of edge e. Graph algorithms developed using templated graphs are generic and reusable software components. DGC (Dual Graph Contraction) has been implemented using templates so that the process can be applied to a wide variety of attributed graphs. Our goals in designing the DGC class were software reuse and readability. We have not optimized the implementation for size or speed, instead slower code which makes the steps of the algorithms obvious has been preferred.

The DGC class encapsulates all the data types and algorithms necessary to conduct multiple levels of dual graph contraction on attributed graphs. The DGC class contains two publicly accessible graphs, *g*, GRAPH<dgc_node, dgc_edge> g, and the dual $\bar{g}$, GRAPH<dgc_dual_node, dgc_dual_edge> dg, both of which are templated to allow the user to provide their own attributes for both the vertices and edges of *g* and $\bar{g}$ by redefining the classes dgc_node, dgc_edge, dgc_dual_node, dgc_dual_edge. DGC also contains two publicly accessible algorithms or methods, contract() which performs one level of contraction and selection() which allows one to interactively define the contraction kernel.

```
class dgc { // Dual Graph Contraction under LEDA
public:
    GRAPH⟨dgc_node,dgc_edge⟩g;
    GRAPH⟨dgc_dual_node,dgc_dual_edge⟩dg;
public:
    void contract(void ); // perform dual graph contraction
    void selection(void ); // interactive selection of the contraction kernel
};
```

The code fragment above shows the public parts of the interface for the DGC class. The graph to be contracted is stored in the templated graph *g* (GRAPH<dgc_node, dgc_edge> g). The classes dgc_node and dgc_edge can be subclassed to store user defined data, a typical usage would be storing the gray values of a pixel. After the user has redefined the attributes of *g* and possibly of the dual $\bar{g}$, the remaining step is to define the decision method used to determine the contraction kernel.

The class method contraction has a default implementation which provides a GUI for interactively defining and saving a contraction kernel. The user selects the oriented edges which should be contracted so that their source node remains. The contraction kernel is stored internally as a list of oriented edges. When replacing this class method the user designs an algorithm which examines the attributes of the nodes, edges of g and possibly $\bar{g}$ and decides upon which edges in g will be contracted and which of their nodes (source or target) will remain.

Once the contraction kernel has been defined the class method contract is called which first contracts edges of g and $\bar{g}$ based on the kernel, and then unnecessary faces, as defined earlier, of $\bar{g}$. In addition contract will construct the dual graph, $\bar{g}$ of g, by calling the class method compute_dual if no dual was supplied.

```
void dgc::contract(void){                    // perform dual graph contraction
    if(dg.number_of_edges()) compute_dual();    // first time compute dual of g
    dual_edge_contraction();
    dual_face_contraction();
}
```

The class method dual_edge_contraction randomly selects and deletes edges from the contraction kernel and calls the class method edge_contract_g to contract that edge of g until the contraction kernel is empty.

```
void dgc::dual_edge_contraction() {
    while(! contraction_kernel→empty()) {
        edge e = ce → choose(); ce → del(e);
        edge_contract_g(g.source(e), e);
    }
}
```

The class method dual_face_contraction is called after contraction in g to eliminate any unnecessary, as previously defined, faces which may have resulted in g. The size of each face of g which is represented by the cardinality of a corresponding node in $\bar{g}$ is examined and any face with degree less than three and not the background face, is added to the set of contraction nodes cn. A single node is selected from this set, and then a single edge from this node is selected to be contracted by edge_contract_dg. Since the topology of g and $\bar{g}$ are changed by this contraction, the algorithm checks if g still contains any unnecessary faces and if so restarts.

```
void dgc::dual_face_contraction(void) {
int change = 0; node_set cn(dg); edge_set ce(dg);
    do {
        change = 0;
        node v; forall_nodes(v, dg) {
            if((dg.outdeg(v)< 3)&&(v ! = BackgroundFace)) cn.insert(v);
        }
        if(! cn.empty()) {
            v = cn.choose(); cn.clear();
            edge e; forall_out_edges(e, v) { ce.insert(e); }
```

```
    if(! ce.empty()) {
        e = ce.choose(); ce.clear(); change++;
        edge_contract_dg(dg.target(e), dg.reverse(e));
    }
  }
} while(change);
}
```

The remaining contraction methods `edge_contract_g` and `edge_contract_dg` are similar and only the later will be discussed. The class method `edge_contract_g` takes two parameters, the edge to be contracted, e, and the node, n of that edge which should remain, currently the node is not necessary since the contraction kernel defines the source node of the edge as the one which will remain. In contraction e is deleted and all edges incident to the target node (in the algorithm cn) of e are moved to n and cn is deleted.

```
void dgc::edge_contract_dg(node v, edge e) {
edge_set oes(dg);
node contract_node = dg.target(e);

    edge_remove_dg(e);        // remove now so e won't be considered in out_edges
    edge oe; forall_out_edges(oe, contract_node) {oes.insert(oe);}
    while(! oes.empty()) {
        edge oe = oes.choose(); edge roe = dg.reversal(oe);
        node target_node = dg.target(oe);
        dg.move_edge(oe, v, target_node);
        dg.move_edge(roe, target_node, v);
        oes.del(oe);
    }
    dg.del_node(contract_node);
}
void dgc::edge_contract_g(node v, edge e) {
                                         // similar to edge contract dual graph
}
```

4. Conclusion

The dgc class is implemented under LEDA and a beta release of the extendable $C++$ class is available which supports dual graph contraction and the saving and editing attributed graphs and contraction kernels with a GUI interface (Fig. 5), as well as animation of the contraction process. The first public release of the system will include image to graph constructors and commented sample applications e.g. watershed segmentation and Voronoi diagram generation, to assist the end user in developing their own applications.

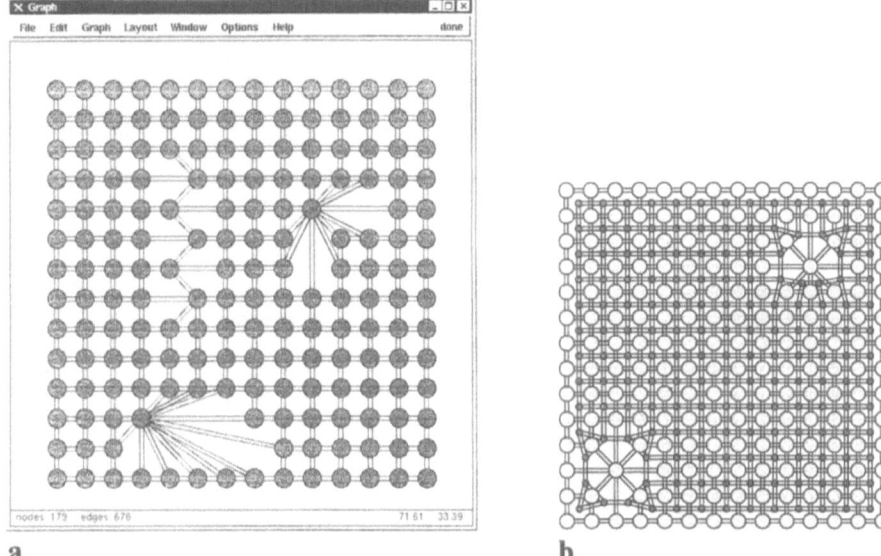

Figure 5. a The GUI for the Dual Graph Contraction tool under LEDA. **b** An example grid graph G with dual $\bar{G}$ shown with smaller nodes. For clarity the background node and it's edges are not shown in the dual graphs

In dual graph contraction, the decimation parameters control the process that iteratively builds an irregular (graph) pyramid, to specify these parameters the concept of the contraction kernel was introduced. It has been shown that dual graph contraction on g preserves connectivity, planarity, and the face degrees of $\bar{g}$. Dual graph contraction provides a general method for specifying segmentations and can be applied to irregular pyramids which are capable of representing all possible segmentations (as defined by Pavlidis [10]) within a single pyramid level [5]. The process may also be applied to regular pyramids, but it has been shown by Bister [1] that there exist segmentation which are not encodable in a single level of such a pyramid.

References

[1] Bister, M., Cornelis, J., Rosenfeld, A.: A critical view of pyramid segmentation algorithms. Pattern Rec. Lett. *11*, 605–617 (1990).
[2] Jolion, J.-M., Montanvert, A.: The adaptive pyramid, a framework for 2D image analysis. Comput. Vision Graphics Image Proc. Image Underst. *55*, 339–348 (1992).
[3] Kropatsch, W. G.: Building irregular pyramids by dual graph contraction. IEE Proc. Vision Image Signal Proc. *142*, 366–374 (1995).
[4] Kropatsch, W. G.: Equivalent contraction kernels and the domain of dual irregular pyramids. Technical Report PRIP-TR-42, Institute f. Automation 183/2, Pattern Recognition and Image Processing, TU Wien, Austria, 1995.
[5] Kropatsch, W. G., Ben Yacoub, S.: A revision of pyramid segmentation. In: 13th International Conference on Pattern Recognition, volume II (Kropatsch, W. G., ed.), pp. 477–481. Washington: IEEE, 1996.

[6] Macho, H., Kropatsch, W. G.: Finding connected components with dual irregular pyramids. In: Visual modules, Proc. of 19th ÖAGM and 1st SDVR Workshop (Solina, F., Kropatsch, W. G., eds.), pp. 313–321. OCG-Schriftenreihe, Österr. Arbeitsgemeinschaft für Mustererkennung, R. Oldenburg, 1995.

[7] Mathieu, C., Magnin, I. E., Baldy-Porcher, C.: Optimal stochastic pyramid: segmentation of MRI data. Proc. Med. Imaging VI: Image Proc. SPIE *1652*, 14–22 (1992).

[8] Meer, P.: Stochastic image pyramids. Comput. Vision, Graphics, Image Proc. *45*, 269–294 (1989).

[9] Mehlhorn, K., Naher, S.: Leda, a platform for combinatorial and geometric computing. Comm. ACM *38*, 96–102 (1995).

[10] Pavlidis, Th.: Structural pattern recognition. New York: Springer, 1977.

W. G. Kropatsch
S. Ben Yacoub
N. Selmaoui
Vienna University of Technology
Institut for Automation 183-2, PRIP Group
Treitlstrasse 3, A-1040 Wien, Austria
e-mail: krw, sby, nazha@prip.tuwien.ac.at

M. Burge
Johannes Kepler University
Department of Systems Science
Computer Vision Laboratory
A-4040 Linz, Austria
e-mail: burge@cast.uni-linz.ac.at

Computing Suppl 12, 111–121 (1998)

Implementing Image Analysis with a Graph-Based Parallel Computing Model

B. Ducourthial, G. Constantinescu, and **A. Mérigot,** Orsay

Abstract

This paper presents a computing model named *associative mesh* and its application for image analysis. The model relies on global computations on subgraphs of an image, as a primitive operation to manipulate objects in an image. While the model is adapted to an efficient hardware implementation, it is also quite expressive in terms of image analysis. The paper illustrates that by presenting the implementation of Voronoï diagrams based image segmentation.

Key words: Parallel computation, image segmentation, Voronoï diagram.

1. Introduction

While graph based image processing is gaining popularity for many applications, its implementation, that relies on dynamically defined data dependent graphs has mainly been considered on serial processors. Indeed, while graphs are powerful tools in terms of expressiveness, their irregular and data dependent nature forbids an efficient implementation with traditional architectures.

This paper presents a new graph based model. Its relies on the definition of subgraphs of the pixels connectivity graph, to describe objects, shapes, etc., in terms of connected components of these subgraphs. The computational primitives that it provides are global operations on these connected sets (global sum, maximum, etc.). This model is suited for a very efficient hardware implementation that will realize these primitives by means of asynchronous data communications [5]. This mechanism insures a fast primitive implementation, while the high level underlying model provides powerful programming facilities for image analysis.

This paper illustrates this formalism in the context of image segmentation, using Voronoï diagrams. Section 2 presents the associative mesh model. Section 3 presents general methods for image analysis using this computing model, while Section 4 is devoted to describe the implementation of Voronoï based image segmentation.

2. The Associative Mesh Model

Given an image with P pixels, we consider a set of P interconnected processors, each assigned a pixel of the image. Let $\mathcal{P}$ be the set of processors, $\mathcal{P} = \{0, \dots, P-1\}$, and $\mathcal{G} = (\mathcal{P}, \mathcal{E})$ the physical interprocessor connection graph, where $\mathcal{E}$ represents the set of edges joining processors in the connection graph. We will assume that $\mathcal{G}$ is a symmetrical directed graph with a constant degree D. For image analysis applications, we shall consider that $\mathcal{G}$ is an 8-connected two dimensional mesh, that is a mesh with diagonal connections.

The *associative mesh model* manipulates two basic objects: parallel variables *pvar* and interprocessor graphs *mgraph*.

The basic object used for communication primitives is a *mgraph*, that is a subgraph of $\mathcal{G}$. More precisely, g is a mgraph iff its set of nodes is $\mathcal{P}$ and its set of edges noted $\mathcal{E}_g$ is a subset of $\mathcal{E}$. Thus, mgraphs are directed (since they are subgraphs of a directed graph). As the set of nodes of a mgraph is implicitly always equivalent to $\mathcal{P}$, a mgraph is only represented on every processor by the subset of the edges of $\mathcal{E}$ that is incident to the processor in the mgraph. The set of mgraphs is noted $\mathcal{M}$.

Connectivity is a fundamental notion for associative nets. We will use the following notations. Given an mgraph g, the *direct ancestors* of processor i in g are noted $\Gamma_g(i) = \{j : j \in \mathcal{P} \text{ and } (j,i) \in \mathcal{E}_g\}$. This definition is extended to define the direct ancestors of any set of nodes $E : \Gamma_g(E) = \bigcup_{i \in E} \Gamma_g(i)$. The *ancestors* of a processor i on a graph g, $\hat{\Gamma}_g(i)$, is the set obtained by the iterated applications of Γ_g, from 0 to $P-1 : \hat{\Gamma}_g(i) = i \cup \Gamma_g(i) \cup \Gamma_g^2(i) \cup \cdots \cup \Gamma_g^{P-1}(i)$. The binary relation R_g defined on $\mathcal{P} \times \mathcal{P}$ by $i R_g j$ iff $i \in \hat{\Gamma}_g(j)$ and $j \in \hat{\Gamma}_g(i)$, is an equivalence relation over $\mathcal{P}$ for mgraph g. An equivalence class is called a *strong connected set* or a *strong connected component*.

As a mgraph is locally coded on every processor, it can be simply manipulated (creation, intersection, etc.) as a pvar, by local data manipulation.

In the associative mesh model, an undirected mgraph can be represented by *symmetrical* edges. It is sometimes useful to return all the edges of a graph. This operation is called *inversion* and the inverse of a mgraph g is noted g^{-1}, $g^{-1} = \{(j,i) : (i,j) \in \mathcal{E}_g\}$.

One of the basic operations that can be performed on a mgraph is an *association*. Given an associative and commutative operator $\otimes$, an undirected (symmetrical) mgraph g and a pvar p, the $\otimes$-*association*(g,p) is a function $(\mathcal{M}, \mathbb{N}^P) \mapsto \mathbb{N}^P$:

$$\otimes\text{-}association(g,p) = \left\{ \begin{array}{l} a_i : 0 \leq i < P \\ \text{and} \\ a_i = \otimes_{j \in \hat{\Gamma}_g(i)} p_j \end{array} \right\}$$

In other words, $\otimes$-*association*(g,p) will return a pvar whose value in vertex i is equal to the iterated application of operator $\otimes$ to pvar p in all the ancestors of i, that is the strong connected set to which belongs vertex i, since the mgraph g is a symmetrical directed graph. Operators we will consider are logical functions (**or**, **and**, **xor**), **max** (maximum), **min** (minimum) and **plus** (addition). As a special case, it is frequently useful to distribute data from a node of the graph to its connected set. This can be done with any *association*, provided the non emitting vertices hold the identity element of the operator.

If the graph g is not symmetrical (most of the time, we consider a tree), we define in a similar way a $\otimes$-*prefix-association*(g, p), that returns in vertex i, the application of operator $\otimes$ to pvar p to all the ancestors of vertex i. When the graph degenerates to a chain of processors, this function is equivalent to parallel prefix computations (scans) [3].

A set of other primitives are also mandatory for the associative net model. In the following, we will use the *spanning-tree* primitive, which is an application of $(\mathcal{M}, \mathbf{N}^P) \mapsto \mathcal{M}$ that, given an undirected mgraph g and a boolean pvar p, will return a forest whose roots will be on processors i such that p_i was initially set. This is done by propagating a market from the roots, with every node randomly selecting one of the first origin of this marker.

Last, we will also consider *associations* restricted to the direct neighborhood of a node, that are called *step-associations*. They are functionally similar to near neighbor operations used in low level image processing, but allow a one step processing on the neighborhood. This operation can combine either a node and its direct neighbors or only the neighbors, in which case it will be called an *exclusive-step-association*.

3. Image Analysis with the Associative Mesh Model

Mesh based parallel architectures are quite popular for image analysis applications [6, 7], as they allow an efficient implementations of basic neighbor-based primitives (convolution, filtering, etc.). The proposed model is a superset of this mesh based scheme as *step-associations* allow, given an adapted graph, to efficiently implement any kind of neighbor based computations. We would like to investigate in this work the interest of the global primitives (*associations*) to solve more complex problems in computer vision, like the image analysis algorithms.

The model allows to deal with image analysis (segmentation, feature extraction, etc.), by using graphs whose connected sets represent objects like regions, contours, etc. Thanks to these graphs, it is for instance possible to either extract basic features on shapes (area, bounding box, etc.) by means of a couple of *associations*, or to realize complex region based operations, as we illustrate in the following. Contour based graphs are formed with frontier pixels. They can be used for shape processing and analysis.

It is also possible to consider graph based regular decomposition of an image. For instance, a graph formed of parallel lines can be used for projections, Hough transforms, etc. With checker like regions, it is possible to realize multi-resolution

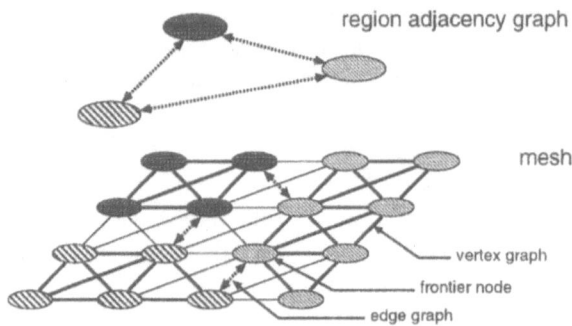

Figure 1. The region adjacency graph and its representation with a virtual graph, composed of an edge graph and a vertex graph

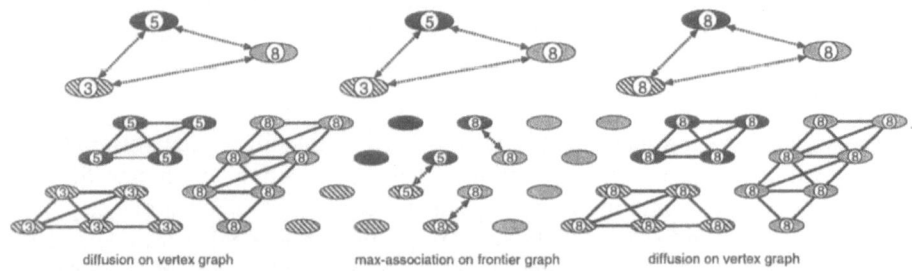

Figure 2. *step–max–association* on the region adjacency graph

processing. Connected sets of the graph are these regions, that can act as a macropixel. Last, directed tree are also useful for problems like distance transforms, contour closing, and so on.

Last, while this model is primary used in the context of pixel based graphs, but it may be useful to consider the so-called *region adjacency graph* for higher level steps of image analysis. This is a graph whose nodes are regions and edges interconnect neighboring regions, and the model can simply be extended to this kind of nonpixel based graphs.

We represented the region adjacency graph by a *virtual graph*, or *vgraph*, whose vertices (regions) are sets of pixels interconnected by a specific graph—the *vertex graph*—and whose edges (inter-region links) are sets of pixels interconnected by the *edge graph* (see Fig. 1). A node that has adjacent edges belonging to the vertex graph and to the edge graph is called a *frontier node*.

This mechanism can be used to embed any graph in the structure [4], without conflicts on processors or links (no congestion and no load) provided the physical machine is large enough.

The associative mesh model can then be used on the virtual graphs. For instance, a *step–association* on these graphs can be done by (see figure 2):

1. distributing data through the vertex graph to all the nodes of the virtual vertices (including frontier nodes).
2. distributing data through the edge graph, in order to diffuse them from frontier nodes of one vertex to frontier nodes of its adjacent vertices.
3. performing a final *association*, by applying the considered operator to the vertex graph. We assume that nodes of the virtual vertices which are not frontier nodes hold the identity element of the operator.

4. Implementation of the Split and Merge Algorithm

4.1. Voronoï Based Image Segmentation

Given a set S of N points in the plane, the Voronoï region associated to $p_i \in S$ is defined by the set of points closer to p_i than to any other point of S. This region will be denoted by $\text{vor}(p_i)$.

Several authors have proposed an adaptation of the *split and merge* image segmentation method by means of Voronoï diagrams (see for instance [1, 9]).

The basic algorithm is the following:

    ▷ *Split the image into regions*
1 Randomly select a set of seed points.
2 Determine the Voronoï region associated to every seed point.
3 Compute an homogeneity criterion associated to the Voronoï region of every seed point.
4 If the region is not sufficiently homogeneous, randomly add a new seed point to the region.
5 Repeat steps 2–4 until all the polygons are homogeneous.
    ▷ *Merge regions:*
6 If a region is surrounded with similar regions, suppress the associated seed.

We will present how this algorithm can be implemented with the associative mesh model.

4.2. Voronoï Regions Determination

The algorithm uses an iterative improvement of regions associated with every seeds. Initial regions are determined by a *spanning-tree* procedure. The initial graph is the physical graph $\mathcal{G}$, and roots of the spanning forest are seed points. If we assume that this procedure is strictly isotropic, a node that belongs to a tree rooted at p_i, will belong to $\text{vor}(p_i)$ and the procedure will directly create Voronoï diagrams. The hardware implementation relies on an asynchronous propagation and this assumption is never verified, but this region is a good initial guess for an iterative implementation. So the algorithm consists to build the regions with the *spanning-tree* procedure (see Fig. 3) and then to improve them to obtain Voronoï regions.

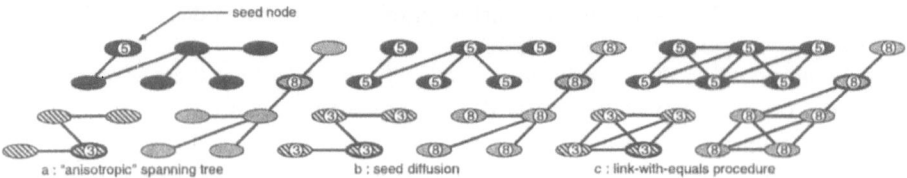

Figure 3. Building initial regions for Voronoï procedure. First, we build a spanning-tree (a). Second, we diffuse the seed index using the tree (b). Then, we link vertices with same value (c), that gives the initial regions

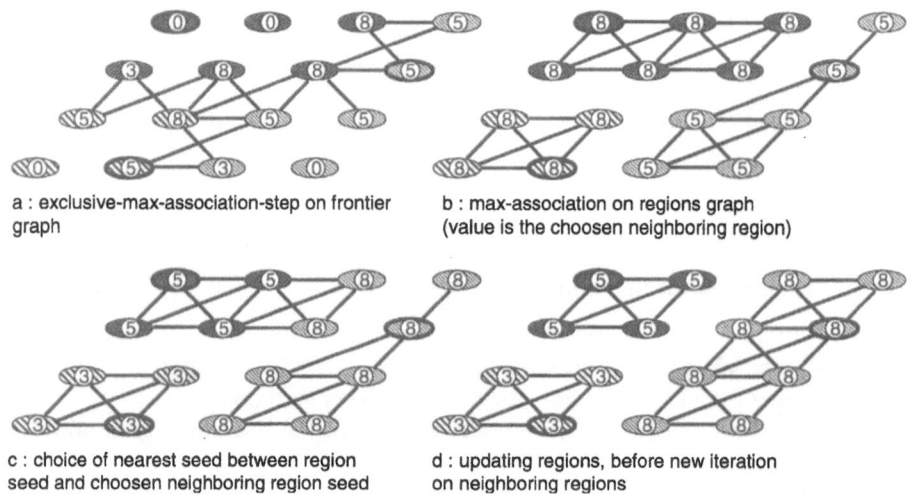

a : exclusive-max-association-step on frontier graph

b : max-association on regions graph (value is the choosen neighboring region)

c : choice of nearest seed between region seed and choosen neighboring region seed

d : updating regions, before new iteration on neighboring regions

Figure 4. Loop on neighboring regions to determine nearest seed (the initial region graph was the one built in Fig. 3)

The iterative improvement of regions is done as follow. The seed coordinates are distributed to every pixel of a region. Then we iterate over the different neighboring regions a diffusion of the coordinates of their seed, and a selection by every pixel of the closest seed. At the end, every point knows what all the closest seeds and we can reconnect the regions. This procedure is iterated until all the regions are stable.

In each loop of the region improvement, all regions iterate a treatment (diffusion and choice of nearest seed for the pixels) over the different neighboring regions. This is done as follows. Each border node selects the largest seed index of neighboring regions, with an exclusive *step–max–association* on the frontier graph (Fig. 4a). Next, each region computes the maximum of the values in its frontier nodes, and then knows its neighboring region with the largest seed index (Fig. 4b). After this neighboring region have been processed – choice of nearest seed and region updating (Figs. 4c and 4d) –, the region suppress all the links from this neighboring region to itself. The process stops when there are no more links between the region and its neighboring regions.

That gives the following algorithm:

   ▷ *Building initial regions:*
1 Build an initial random forest rooted at seed nodes with the *spanning–tree* primitive. Let Forest be this mgraph.
2 Distribute through Forest the coordinates of the seeds (root nodes). Let Seed be this value.
   ▷ *Loop until stability of regions:*
3 Repeat forever
4   Build the regions graph linking all pixels with the same value (that is the same seed). Let Region be this mgraph.
   ▷ *The frontier graph is the complementary of the region graph in the mesh. Let* Frontier *be this mgraph.*
5   Set Frontier = $\overline{\text{Region}}$.
   ▷ *Loop on neighboring regions.*
6   While Frontier is not empty
7   Frontier nodes compute the maximum index seed through mgraph Frontier with a *step–max–association*(Frontier, Seed). Let m be this value.
8   Compute with a **max–association**(Region, m) the global maximum of m on mgraph Region. Let M be this value.
   ▷ M *is the seed of the neighboring region examined in this iteration.*
9   Every node computes its distance to node of index M. Where this distance is smaller than the distance to Seed, NewSeed = M, elsewhere New Seed = Seed.
10   Frontier nodes suppress edges in mgraph Frontier that lead to the region rooted at M.
11   End while
   ▷ *Now every node knows the index* NewSeed *of its closest seed point. We just update* Seed. *If there is no changes, it is the end of the algorithm.*
12   If Seed == NewSeed in every node then exit loop.
13   Set Seed = NewSeed
14 End repeat

4.3. Computing the Homogeneity Criterion

Computing the homogeneity criterion of a region R can simply be done with the associative net model. Currently used criterions are: variance$(R) < \varepsilon$ or $\mathbf{max}(R) - \mathbf{min}(R) < \varepsilon$, frequently associated to an area measurement to avoid to split small regions. All these computations just require a couple of global *associations* on the regions (connected sets of the region graph). For instance, the $\mathbf{max} - \mathbf{min}$ computation on regions described by mgraph g, is just **max**-*association*$(g$, grey-level)— **min**—*association*$(g$, grey-level).

4.4. Adding a New Seed Point

In general, adding a new seed point is done on a random basis. For instance, we can generate random numbers on the images with any method and select the pixel with the largest number with a **max**-*association*. To avoid multiple selections, we simply append the pixel indexes before the maximum computation to insure uniqueness.

The model also allows to use simple heuristics to improve the generation. For instance, the next algorithm shows how to split the image in 8 octants, and to add a point at the centroïd of the octant whose average grey level is the most different from the average grey level of the initial region.

   ▷ *Define an octal partition of the region centered at seed points.*
1 Mark neighbors of the seed points with a *step–or–association*.
2 Span a forest rooted at the previously marked 8 neighbors of the seed point. This defines an octal partition of the region.
3 Compute on these partitions their average grey level by a **plus**-*association* on the trees of this forest. Let AGL be this value.
4 Compute the difference between initial value and AGL. Let D be this value.
5 Select the partition with the largest difference to the average grey level of the whole region with a **max**-*association* on D in the complete region graph.
6 With a couple of **plus**-*associations*, compute on the selected partition its geometric centroïd.
7 Select the point that is the closest to this centroïd (with a **min**-*association*) and add a new seed at that point.

4.5. Region Merging

Region merging implies to decrease the number of regions by removing regions, provided the homogeneity criterion is respected. More precisely, a region R will be suppressed if *all* its neighboring regions are sufficiently similar to it, in terms of homogeneity criterion. This will insure that the surrounding regions, whilst augmented by parts of R, can still be coded by a Voronoï scheme with the same seeds. To realize this operation, we will use the vgraph approach presented in Section 3. We assume that vertices of the vgraph hold all the informations required to compute the homogeneity criterion (for instance maximum and minimum grey levels for the **max** − **min** criterion, or the sum of the grey levels and the region area for criterion based on the average grey level, etc.). For sake of simplicity, we will assume later that the criterion is **max** − **min**. In that case the homogeneity of a set of regions is just computed by a *step*–**max**–*association* of the region maxima on the region vgraph, and a *step*–**min**–*association* of the region minima on the same vgraph.

An additional problem arises from the concurrency of the fusions, that can lead to an incorrect merging. This is solved by forbidding the simultaneous suppression of neighboring seeds. This concurrent merging control can be done by forming a graph connecting neighbor regions that are to be suppressed, and by keeping a unique node per connected sets.

The corresponding algorithm is described below:

1 Let RAG be the region adjacency vgraph, max and min the maximum and minimum values in each region, and Seed a boolean pvar indicating if a node is a Voronoï seed.

2 Compute MAX = *step–max–association*(RAG, max).

3 Compute MIN = *step–min–association*(RAG, min).
   ▷ MAX − MIN *gives the amplitude of the region formed by merging every node with its neighboring regions.*

4 Let Homogeneous be true if MAX − MIN < threshold.

5 Where Homogeneous is true
   ▷ *We must now avoid concurrent merging of neighboring regions.*

6   Create a vgraph connecting a node to its neighbors where Homogeneous is set.
   ▷ *Keep only one node per connected sets of this graph.*

7   Compute the **max–*association*** of node indexes on this vgraph. Let Selected be this value.

8   Where Selected ≠ index
   ▷ *There is an homogeneous neighbor and it has been selected. Do not merge at the present step.*

9   Set Homogeneous to false.

10   ElseWhere

11   Set Seed to false in order to suppress the seed and its region.

12   End Where

13 End Where

14 Recompute the Voronoï regions with Seed indicating the seed points, and build the region adjacency graph.

15 Iterate 1–13 until no seed point can merge any longer.

5. The Simulation Library

A library to emulate the behavior of associative meshes has been written (*sam: simulate the associative mesh*). It is based on C_{++} classes that represent the basic objects manipulated by the model. Image classes allow to perform data parallel point-wise manipulation of images à la C⋆ [8]. Graph classes can also be created and manipulated in a data parallel way. They are used for local (*step–associations*) and global (*associations*) transformations on images. Operation domains can be restricted to a subset of images, thanks to parallel-if constructs named *where* (and *elsewhere*).

A Voronoï segmentation has been implemented on top of this library. As an example, a coding fragment of the main loop of the program is presented below.

```
//Fragment of the Voronoï procedure coded with the sam library
//(for clarity, pvars are prefixed with pv·and mgraphs are prefixed with mg)
do{
   //mgRegion is the mgraph of regions. pvSeed is the pvar of region indexes.
   mgRegion.LinkWithEquals(pvSeed, pvSeed)//Building of mgRegion.
   //mgFrontier is the frontier graph.
   mgFrontier = ~mgRegion;           //Building of mgFrontier from mgRegion.
   //pvDistSeed is the pvar of distance from the seed of the region.
   pvDistSeed = SeedDistance(pvSeed);       //Computing pvDistSeed.
   //Loop on neighboring regions.
   do{
      //The where condition is true if the region has a neighboring region.
      WHERE(OrAssoc(mgRegion, mgFrontier)){
         //pvNeighborSeed is a pvar containing the index of the chosen region.
         pvNeighborSeed = MaxAssocStepExclusive(mgFrontier, pvSeed);
         pvNeighborSeed = MaxAssoc(mgRegion, pvNeighborSeed);
         //Compute pvDistNeighborSeed, the distance from pixel to distance from
         neighbor region seed.
         pvDistNeighborSeed = SeedDistance(pvNeighborSeed);
         WHERE(pvDistSeed > pvDistNeighborSeed)
            pvNewSeed = pvNeighborSeed;//Update pvNewSeed
         ELSEWHERE
            pvNewSeed = pvSeed;
         ENDWHERE;
         //Suppress links from region # pvNeighborSeed to region # pvNeighbor,
         using a temporary mgraph mgTemp.
         mgTemp.LinkWithEquals(pvNeighborSeed, pvSeed);
         mgFrontier &= ~mgTemp;
      }ENDWHERE;
      //mgFrontier.GlobalOr( ) = 0 if it has no edge.
   }while(mgFrontier.GlobalOr( )!= 0);
   //Test of any change in the region. pvEnd is used in the while condition below.
   WHERE(pvSeed != pvNewSeed)
      pvEnd = 1;
   ELSEWHERE
      pvEnd = 0;
   ENDWHERE;
   pvSeed = pvNewSeed;           //Update pvSeed
   //pvEnd.GlobalOr( ) = 0 if there is no change in the regions.
}while(pvEnd.GlobalOr( )! = 0);
```

6. Conclusion

We have presented in this paper the use of an high level model for graph based
image manipulation. This model can present a double benefit. First, as it allows

an efficient hardware implementation, it can be used to build a fast image analysis computer. Second, at the software level, the model can be the basis of an interactive image processing environment. It has for instance be integrated in the PACCO programming environment, where it allows to develop image analysis by means of a scripting language [2].

References

[1] Ahuja, N., An, B., Shachte, B.: Image representation using voronoi tesselation. CVGIP *29*, 286–295 (1985).
[2] Biancardi, A., Mérigot, A.: Connected component support for image analysis programs. Proc. 13th ICPR, *IEEE Press*, D:620–624, 1996.
[3] Blelloch, E. G.: Vector models for data-parallel computing. Cambridge Mass.: MIT Press, 1990.
[4] Ducourthial, B., Mérigot, A.: Graph embedding in the associative mesh model. Technical Report 2, Institut d'Electronique Fondamentale, Bât. 220. Université Paris-Sud, 91405 Orsay, France, December 1996.
[5] Dulac, D., Mohammadi, S., Mérigot, A.: Implementation and evaluation of a parallel architecture using asynchronous communications. In: Proceedings of the 1995 Workshop on Computer Architecture for Machine Perception *(CAMP'95)* (Cantoni, V., Lombardi, L., et al., eds.), pp. 106–111. Los Alamitos: IEEE Press, 1995.
[6] Fountain, T. J.: Processors arrays: architecture and applications. New York: Academic Press, 1987.
[7] Owens, R. M., Irwin, M. J., Nagendra, C., Bajwa, R. S.: Computer vision on MGAP. In: Proceedings of the 1993 Workshop on Computer Architecture for Machine Perception (Davis, L. S., Bayoumi, M. A., Valavanis, U. P., eds.), pp. 337–341. Los Alamitos: IEEE Press, 1993.
[8] Rose. J. R.: C*: C++ like language for data-parallel computation. Technical Report PL87-8, Thinking Machines Corporation, Cambridge, Massachusetts, USA, 1987.
[9] Tuceryan M., Jain, A. K.: Texture segmentation using Voronoi polygons. IEEE Trans. PAMI *13*, 211–216 (1990).

B. Ducourthial
G. Constantinescu
A. Mérigot
Institut d'Electronique Fondamentale
Université Paris sud
91405 Orsay
France
e-mail: {ducourth, constant, merigot}@ief.u-psud.fr

Computing Suppl 12, 123–134 (1998)

The Frontier-Region Graph

J.-G. Pailloncy and **J. M. Jolion,** Villeurbanne

Abstract

In this paper an image representation is presented, suitable for multithreaded process and pyramid. A memory requirement comparison and an efficient, fast & small algorithm to build the graph is presented.

Key words: Region adjacency graph, hierarchical graph, pyramid, multithreaded process.

1. Introduction

In this paper, we will present some theoretical developments about graphs as image representations. Graphs have been used in the image analysis fields for many years but still not have a complete understanding of what should be their appropriate properties for image manipulation purposes. Indeed, the older graph representations retain some defaults. For instance, the Region Adjacency Graph (RAG) [5, 8] misses some features like frontiers' number, inclusion, unique representation, connectivity paradox [3, 7], or multi-threaded and parallel process.

Our main goal is to design an efficient graph representation of an image, suitable for pyramid merging. An image's pyramid can be associated to a region merging process; the base of the pyramid being the input image, a given level, an intermediate state of the region growing, the apex, the final segmentation (obtained from a given set of parameters). We want to be able to flip from a segmentation to an other (i.e., characterized by another set of parameters) without having to compute again the whole pyramid. We so need a compact and efficient structure to implement the pyramid. We use a labeled graph, where the graph represents the first segmentation, and the label the merging pyramid.

The *frontier graph* may be built on the cell complex theory from Kovalesky. We introduce a smaller representation of the frontier graph with the same properties. Our contribution will be to show that the memory requirement of the *frontier graph* is *linear with the size of the input image*, and most basic functions to access the data are in constant time, that *every element* in the structure are *constant in size*, and that the *frontiers* are the *constant objects of the graph*, and that the *transitive closure* feature is inherent to the structure.

2. Topological Representation

We want to use the graph representation of the image for segmentation purposes. An irregular pyramid is built on the graph representation [2]. We use the cell complex to describe an image (Fig. 1), because it is topologically consistent. In addition, this frameworks allows us to apply any results to any tessellations. An other approach may be based on the star topology [1].

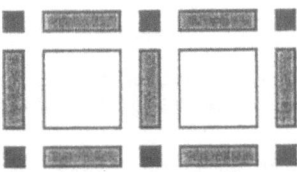

Figure 1. A classic image, a regular 2D array of square pixel, we have black points, grey linels, white pixels, and the classic bounding relationship. A linel is equivalent to the couple of adjacent pixels, point is equivalent to the 4 adjacent linels

We will first compute the size of the graph. The size of the canonical cell complex [3, 4] associated to a nD regular torus of pixels, with N the pixel array's size, is:

$E = (n2^{n-1} + 1)N$ for the number of elements

$B = (n(n + 1)2^{n-2} + 1)N$ for the number of bounding relationship

In 2D, $E = 5N$ and $B = 7N$.

Let m be the average size of a region at the first level of the pyramid. We thus have at the first level $O(N/m)$ regions. The size rate between the canonical cell complex graph and the region ones is $O(m(n2^{n-1} + 1))$. In 2D, with an average size of 10 pixels, the number of node of the region's graph is 50 times smaller than the cell complex ones. We want to reduce the memory footprint of the image. Indeed, this naive structure is not very useful as a memory representation of the image: the related graph is too huge. But it gives us a correct description of the image (topology). Moreover the notion of region as a connected area of an image is not explicit in this representation. We have a full and correct description of the pixels, but we do not have a suited description of a set of pixels.

3. The Frontier Graph

We now want to debate which structure is the most adequate for our work: to represent the structure of the Frontier Graph in memory. The first thing that we need, is the frontier of each region. It allows us to compute any other properties of the regions. We will consider different structures like the naive one, the frontier graph or the array one.

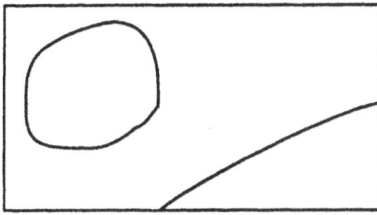 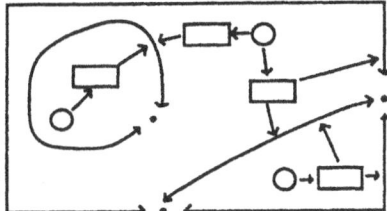

Figure 2. Any graph's element is represented by a structure: the region (circle), the sets of frontiers
(square), the frontier (edges), and the full duplex link between each couple of object

3.1. The Naive Structure

The naive structure define, one for one, each element of a cell complex in
memory.

For each level of the pyramid we have a new complete structure like the one in
Fig. 2. This representation was fully developed and used on both a SUN 20 and a
CM5 (parallel computer with 32 nodes) [6]. The weak point is the memory footprint. For a gray image 800*800 with 30.000 regions at the first level of the
pyramid, and 400 levels in the pyramid, we need more than 600 Mbytes of memory. It was a very simple structure, but useful to test the theory.

3.2. Frontier

The frontier's graph just describes the frontier, but gives enough information to
compute the regions. Each edge of the graph (*i.e.*, frontier), at each of it's end,
will point to the next edge. In addition, each edge will get a mark to know the
direction of the edge. The mark will be the regions' IDs of the two adjacent
regions. Because each edge points to the next one, the edges' list at a given vortex
is ordered.

The graph is planar. It exists an edge between two regions, and on each vortex the
adjacent edges are ordered. All these properties allow us to build the whole graph
except for a direct continuous transformation. We have a complete geometrical
description for each edge, which allows us to build the image tessellation.

As a classic process, we want to be able to follow the contour of a region. The
edge is oriented and we associate a one-bit flag to each edge's link to know from
which side we arrive on it. An other solution may associate a pointer to the four
neighbors to each edge, and not only to two ones. In Fig. 3 (right), suppose we
arrive on the edge A from C, the next edge around the region R is B or B'. The
on-bit flag shows that we arrive from the origin of the oriented edge A, so we
know now that the next edge is B (whatever is the orientation of B).

In what follows, all the sizes will be in bytes. Let N_f be the number of frontiers in
the graph, w the pointer's size, c_f the frontier's characteristic size. A frontier is

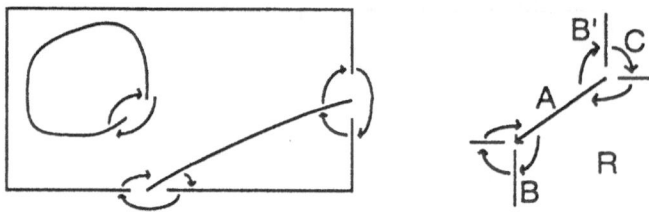

Figure 3. Simple Frontier Graph and on-bit flag example for frontier following

described by a characteristic structure, a level (w) and two pointers to the next frontiers, and at least the direction (on 2 bits). The frontier size is bigger than $c_f + 1/4 + 3w$. Most of the time we include two pointers to the adjacent regions, so the size is $c_f + 5w$. The total size for the pyramid is at least $(c_f + 1/4 + 3w)N_f$, and mainly about $(c_f + 5w)N_f$.

c_f will depend on the length of the frontier. With a regular squared pixel tessellation, we code the frontier with the Freeman code. In this code, we describe the frontier by one of its ends followed by the list of direction changes. The size is size_of(point + first direction) + length(frontier)/5. If we do not need the absolute position of the frontier in the image we just have the size equal to the length divided by 5. Indeed, it is possible to code 5 direction changes in 4-connected paths (i.e., 3 different direction changes) in 1 Byte ($3^5/256 = 94$, 92%). This is one of the smallest and simplest solutions.

3.3. Array

The smart point of the array structure is its tessellation regularity to speed up the process and reduce the memory size, but this applies only on a regular tessellation. If we have a number of region near equal to the pixel's ones at the first level of the pyramid, we can implement the frontier's graph by an array. Let be a 2D image of square pixels, with a size of $N = n*m$, we have $(n-1)m + n(m-1) \cong 2N$ frontiers. We need a structure of $2N*$size_of(boolean) $= N/4$ bytes to code the frontiers for one level, we need $\cong 2N*$size_of(level) $= 2N*w$ bytes to code the whole pyramid.

3.4. Comparison

3.4.1. Complexity

We will compare the requirement of the different structures and find the limit of usefulness for each. An important result is that the size's complexity is linear with the size of the image. The region with a frontier's list is the only structure that does not have a linear memory requirement, but $O(N_f N_r)$. And it is possible to have most of the basic functions computed in constant time.

	Smallest	Frontier	Array
Frontier's size	$3w + 1/4 + c_f$	$5w + c_f$ ˙	w
Frontier's number	N_f	N_f	N
Total	$(3w + 1/4 + c_f)N + c_r N_r$	$(5w + c_f)N_f + c_r N_r$	$2wN + c_r N_r$

The smallest graph is the smallest frontier graph in size.
c_r, c_f Region, Frontier's Characteristics
w Computer word (2 or 4 bytes)
N_f, N_r Number of frontiers $(N_f < 2N)$, regions $(N_r < N)$
N Number of tesselles (Pixels)

We now study the factor of the linear complexity, and compare the memory requirement of each case. Let a, b be such as $N_f = 2aN$, $N_r = bN$. We have $0 < a, b \leq 1$. $N_r \leq N_f$ implies $b \leq 2a$. $N - N_r \leq 2N - N_f$ implies $b \geq 2a - 1$. We will compare further the smallest (S), the frontier (F) and the array (A) structure's sizes:

$$S = (3w + 1/4 + c_f)N_f + c_r N_r = [(3w + 1/4 + c_f)2a + c_r b]N$$

$$F = (5w + c_f)N_f + (c_r + k)N_r = [(5w + c_f)2a + (c_r + k)b]N$$

$$A = 2wN + (c_r + k)N_r \qquad = [2w + (c_r + k)b]N$$

where k represents the region's size supplement if we use the explicit region. Implicit regions are not represented in memory: a particular frontier knowed as describing a region is enough to represent the region. On the other side, an explicit region is represented in memory by a structure coding theirs full properties.

$k = 0$ (implicit), $k = 6w + c_r$ (Region)
or $k = 6w + c_r + \langle$Number of adjacent frontiers$\rangle$ (List)

Moreover, c_f may not be constant. It may depend on the length of the frontier, here $c_f = c_f^0 + (1 + (2N - N_f)/N_f)e = c_f^0 + e/a$, and $e = 1/5$.

3.4.2. Ratio of the regions' size

The most interesting comparison is to know when the array is more appropriate or not than the graph to implement the frontier's graph for a given set of parameters. We will study the graph versus the array, the smallest graph versus the array.

The condition's limit between the graph and the array is given by:

$$F = A \quad \text{or} \quad (5w + c_f)2a + (c_r + k)b = 2w + (c_r + k)b \quad \text{or} \quad (5w + c_f)a = w$$

Consider a classic case where indexes code using two bytes $(w = 2)$, Frontier's characteristic is $c_f = c_f^0 + e/a$, $c_f^0 = 6$ and $e = 1/5$, we thus have $a = 9/80$, $1/a = 8.889$. The condition's limit between the smallest graph and the array is given by:

$$S = A \quad \text{or} \quad (4w + 1/4 + c_f)2a + c_r b = 2w + (c_r + k)b$$

$$\text{or} \quad (4w + 1/4 + c_f)2a = 2w + kb$$

We note that the region's type is now important. Consider the same case with $k = 0$, we get $a = 72/285$, $1/a = 7.916$. It's normal 7.916 is smaller than 8.889, the smallest graph structure uses less memory than the standard one.

So, the graph structure is more interesting to use than the array structure to code the frontier's graph when the average size of the region at the first level of the pyramid is greater than 8.889 pixels. When we segment an image, we want to reduce the number of region, the region's average size of the pyramid's first level is thus often greater than 10. In the precedent case and commonly we have $1/a < 10$, we may so use the graph structure instead the array one.

4. Region

There are different ways to describe a region: as a frontier, as a list of frontiers, or as a set of two frontier-regions. And we may mix the different representations.

4.1. Characteristics

Whatever is the representation of the region, the characteristics may be stored or computed. There are two region types. For the first level of the pyramid, we need the basic characteristics of the region: its shape, its color. If the tessellation is regular, the shape is constant and does not need to be stored. The second ones are the regions inside the pyramid, which may compute theirs characteristics each time we need them, or only once when they are created, or as a compromise each time we need them but with a buffer mechanism.

4.2. Structure

The frontier graph gives us enough information about frontier, we now want to study the region's representation.

Consider the naïve structure where we keep an adjacent frontiers' list for each regions (Fig. 4, left). Let N_f be the number of frontiers, and N_r the number of regions of the first level, the size is $O(N_f N_r)$ (*i.e.* not linear). It is possible to represent a region by a double word. The first one is the frontier's ID of one of its frontier, and the second one is the level of the region. And we identify the region's pointer to this record (double word). If we choose in addition the frontier's ID to have the same level ID as the region one (this is always possible) we only need one word to describe a region. This point is of importance: there is no difference

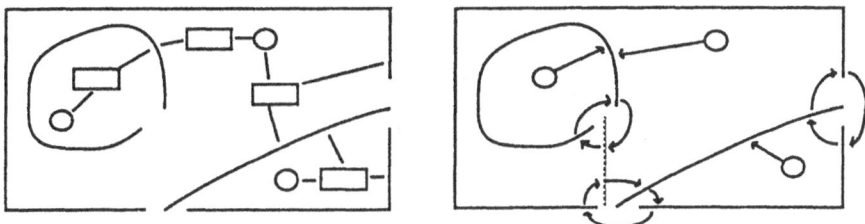

Figure 4. Left: The naive representation has to describe each region (circle) by the full list (square) of its adjacent frontiers. Right: Each region point to an adjacent frontier, a fictive edge connect the hole with its surrounding region

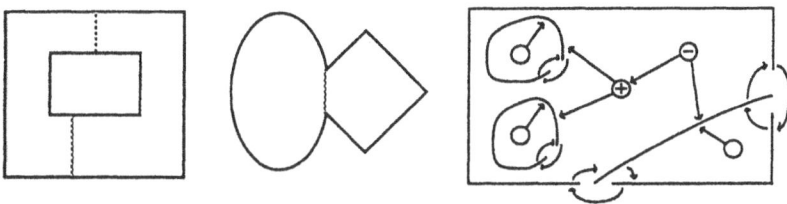

Figure 5. Left and middle: region with hole and overlapping regions with fictive egdes. Right: boolean representation of region's inclusion

between a frontier and a region. A region is just a frontier known as a region (Fig. 4, right).

The weak point is the region with a hole. We restrict ourselves to a simply connected partition of the space. This constraint affects only the first level of the pyramid. Any other level could have non simply connected region. We build a fictive level with a label −1, and split the region in simply connected subregions. It is very useful to be able to build a level −1. Indeed, if we have an image where two real objects look like only one region (Fig. 5), we need to be able to split a region at any pyramid's level. As we need this property, we may use it to describe a tesselle with the hole as a set of simply connected sub-tesselles. The first level of the pyramid must not be a simply connected tessellation, but the lowest level must do.

An other solution is to represent a region as a simply connected region, or as a Boolean operation between two other regions (Fig. 5, right). A region with a hole is the simply connected surrounding region deprived of the included region. Here the two included regions are merged and after subtracted to the surrounding one. It is always possible to choose a merging order so that the adjacency of two merged regions is preserved.

4.3. Comparison

We will compare the performances of the different structures. The irregular pyramid's algorithm builds a level zero, which is the initial partition of the image. A new level is built by merging the neighbor regions having the same characteristics. For each couple of regions that are candidates for merging, most of the criteria require the characteristics of the common frontier (Characteristics, Level), the characteristics of the adjacent regions, the list of the frontiers bounding these two regions (First/Next Bounded/Bounding Object), and the father and sons of any regions (Father/Sons) in the pyramid graph. More, we need a mechanism to lock the access to each object (Lock), and a way to know for each object if it has been merged or not.

If we do not use explicit regions only the functions shown below will not have a constant time. Using a region to store the characteristics speedups the process, but requires more memory.

	Smallest	Frontier	Array	Region
Get Father/Sons	$\#adj.f.$	$\#adj.f.$	$\#adj.f.$	Constant
Set Lock	$\#adj.f.$	Constant	Constant	Constant
Get Region Characterictics	Heavy, C_r	Heavy, C_r	Heavy, C_r	Constant
Set Region Characterictics	N.A.	N.A.	N.A.	Heavy, C_r

N.A. Non-existing operation, $\#adj.f.$ Number of adjacent regions.

We may buffer the computed characteristics and the son's IDs during the pyramid building. This requiers half less memory. We will use the frontier or array structure for the pyramid with a region characteristics buffer. The only slow functions are to get the father. During the building of the pyramid we sparely need it.

5. Pyramid

5.1. Structure

The irregular pyramid, represented as a labelled graph, is a part of a progressive image representation. Under the pyramid we have the pixel array: perfect for measurement but without any structure information. At the pyramid base we have the regions defined by their borders: more structure and less measurement. The Frontier Graph and the pyramid extend the structure with the neighbourg and the father-son relationships and loose the distance measurement. We may introduce an object concept to represent not only a region (connexe image part) but a region as a sub-graph (i.e., a region with the labelled frontiers' hierarchy). In future, we will have an other source of knowledge like a stereo image, a knowledge database, an image sequence, we will extend the structure by introducing a new element. For example, we may use an object to represent a tracked region over an image sequence.

5.2. *Algorithm*

We now present the general algorithm to build an irregular pyramid defined with the frontier's graph, and detail its time complexity. There are two steps to build the irregular pyramid.

5.2.1. Creation of the Base Level

From an input image, we want to build the base level of the pyramid. The first step is to filter and label the image. The filter allows us to reduce the number of regions, and gives us the region's characteristics (which otherwise may be one pixel most of the time). Now we consider we have as an input a connected component labeled image, and we assume each first pixel's region are known.

The following-frontier algorithm is easy and fast: we just follow the frontier until we discover a new one, and we add it to the graph. The time complexity is O(length of the frontiers). If we do not have the list of the frontiers' first pixels, we need to scan the interior of the regions to find the holes. The complexity will be $O(N)$.

The second part of the algorithm is to connect holes with their surrounding regions, if any. The maximum number of frontiers is obtained when all the frontiers between the pixels become a frontier of the graph. And there are at least 4 pixel's frontier less for each hole in the image. We thus get the time complexity $O(N_f + \text{number of holes}) < O(\text{number of pixels' frontier})$. The time complexity is $O(N)$ in the worst case.

5.2.2. Merging Process

Given a frontier graph, building a pyramid is just to label the edges (*i.e.*, frontiers) of the graph. The level i of the pyramid is the subgraph where the edge's labels are i. A labeled graph is a reduced form of the pyramid. It is efficient because the frontiers are not modified during the merging process.

We can not detail a general algorithm because it too much depends on the properties of the merging process: the frontiers' evaluation depends on their neighboring. Commonly the evaluation of a frontier depends on its characteristics, and on the characteristics of its two adjacent regions. Each time a frontier is merged, we need to update each frontier adjacent to all the regions surrounding the new one. At each level of the pyramid, we need to compute again quite all the frontiers' evaluations.

If we limit ourselves to merge only regions' pairs, we may use the following three-pass algorithm. At first, each frontier is evaluated (E) independently to the others. On a second time, for each region we choose the frontier that will be merged (C). The third pass is the merging process (M), where we just set the frontier and

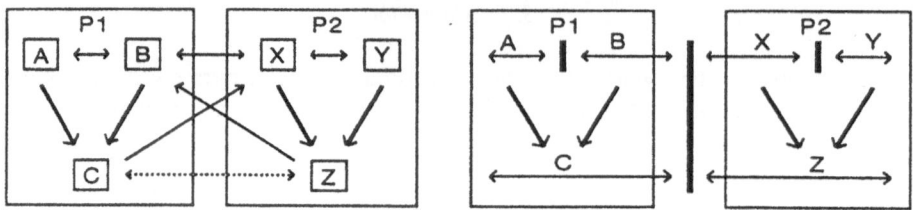

Figure 6. Multi-threaded merge process with RAG and frontier graph

region's level, and where we compute the regions' characteristics if necessary. Each pass has a time complexity $O(N_f)$.

The creation time complexity of the pyramid's base is always linear in the number of pixels. Theoritically the building of the whole pyramid may be linear in the frontier's number, if the frontier's evaluation is independant of the precedent levels. In the previous case the creation of a new level is linear in the frontier's number.

5.3. Coherence

We now study the multi-threaded approach to the merging process. Consider the Region Adjacency Graph (RAG), when merge a region we need to maintain all links to it. In Fig. 6 (left), the thread $P1$ want o merge the regions A and B in C and in the same time the thread $P2$ the regions X and Y in Z. After the merging process, the region C has a pointer to the region X inherited from B but the neighbor of X is Z. To maintain the coherence, we may after the merging process walk through the pyramid to find the new neighbor. The thread $P1$ looks for X and X's father, then for Z, and so on.

The RAG is not so useful for multi-threaded process. The reason is that the RAG does not represent the constant (non modified) objects of a pyramid. When we merge two regions, the regions change, but the frontiers stay the same. The only frontier, from which we have merged, just disappears.

Consider now the frontier graph. On the same example Fig. 6 (right) the regions are represented by the links between frontiers. The regions A and B are merged in C. But the frontier between B and X is not modified. There is no need to maintain the coherence.

5.4. Parallel Computer

A solution to the intensive computation operation is to use a parallel computer. This technology allows us to easily expand the computation power, but the main troubles are the coherence, the communications and the work's load. The frontier graph reduces the coherence and communication's troubles. The coherence property developed for mutli-threaded process also applies.

Any objects have a constant size, and most of the data are constant. During the merging process, the only modified objects are the labels. Because the exact number of objects is known at the beginning (as a result of the filtering process), all memory allocations are also done at the beginning. In the naive representation [6] the memory allocation represents 15% of the computation time. The size of the labels is wN_f, which is less than $1/10$ of the whole structure size $((5w + c_f)N_f + (c_r + k)N_r)$. We can duplicate not only the label table but the whole structure on all processors. For a 1024^2 image on a 32 bits computer $(w = 4)$ with a region's average size 10 at the first level $(m = 10)$, the memory footprint for the labels is only 410 Kbytes. The total size is 6.6 Mbytes for a classic structure $(c_f = 6, c_r = 8, N_f = N_r = N/10, k = 6w + c_r)$.

The communication process depends too much on the frontier's characteristics computation to be developped here. We study the case where all the data are duplicated on all processors. Let p be the number of processors. We suppose that the load is balanced at the beginning. During the communication of the frontier's evaluation (E), the data size (by level) that the processors must broadcast is $4w(pN_f)^{1/2}$. For a 1024^2 image with $w = 4, m = 10$ on 64 processors, this is 60 Kbytes. During the communication of the chosen frontiers (C), the total broadcasted data size (for all the levels) is wN_f. For the same image, this is 409 Kbytes. During the communication of the region's characteristics, if any, the total broadcasted data size (for all the levels) is c_rN_r. For the same image with $c_r = 6$, this is 307 Kbytes.

There is a communication part independent of the level's number in the pyramid that is $wN_f + c_rN_r$ and an other part that is $4w(pN_f)^{1/2}$ at each level.

The frontier graph is simpler than the RAG on parallel computer. It easily maintains the coherence across processors. The amount of data transferred during the merging process is small. All the memory allocation and the partition data are done at the beginning.

6. Future

We have a representation of a graphs' pyramid with a linear requirement in size and time. One of the future direction will be a 3D extension. The cell complex allows us to easily extend the representation to an upper dimension in theory, but the memory representation of the surface's structure become more complex. An other direction: A frontier graph may code different segmentations in only one structure (to save some memory requirements). Generally, we will study the operators to handle the pyramid, and deeply analyse the parallel algorithm in some real cases.

References

[1] Fioro, C.: The topologically consistent representation for image analysis: the frontiers topological graph. Discrete Geometry for Computer Imagery, 6th International Workshop, Lyon, France, Nov. 1997.
[2] Jolion, J.-M., Montanvert, A.: The adaptive pyramid: a framework for 2D image analysis. Comp. Graph. Image Proc.: Image Und. *55*, 339–348 (1992).
[3] Kovalesky, V. A.: Finite topology as applied to image analysis. Comp. Graph. Image Proc. *46*, 141–161 (1989).
[4] Kovalesky, V. A.: Digital geometry based on the topology of abstract cell complex. Geom. Discr. Imag. 259–284 (1993).
[5] Kropatsch, W.: Equivalent contraction kernels and the domain of dual irregular pyramids. PRIP-TR-42, Technical University of Vienna, 13 Nov. 1995.
[6] Pailoncy, J.-G.: Classification d'Images satellites sur CM5 (Rapport de DEA, Juil. 1994).
[7] Pavlidis, T.: Structural pattern recognition. New York: Springer 1977.
[8] Rosenfeld, A.: Adjacency in digital pictures. Inform. Control *26*, 24–33 (1974).
[9] Zucker, S. W.: Region growing: childhood and adolescence. Comp. Graph. Image Proc. *5*, 382–399 (1976).

J.-G. Pailloncy
J.-M. Jolion
Laboratoire Reconnaissance de Formes et Vision,
Bât. 403/INSA
F-69621 Villeurbanne Cedex, France
e-mail: jgpaillo, jolion@rfv.insa-lyon.fr

Computing Suppl 12, 135–145 (1998)

© Springer-Verlag 1998

Optimization Techniques on Pixel Neighborhood Graphs for Image Processing

V. V. Mottl, A. B. Blinov, A. V. Kopylov, and A. A. Kostin, Tula

Abstract

A class of image processing problems is considered from the standpoint of treating them as those of co-ordinating the local image-dependent information and a priori smoothness constraints. Such a generalized problem is set as the formal problem of minimization of a separable objective function defined on an appropriate pixel neighborhood graph. For attaining a higher computation speed, the full pixel lattice is replaced by a succession of partial identical neighborhood trees. Two versions of a high-speed minimization procedure are proposed for, respectively, discretely defined and quadratic objective functions.

Key words: Image processing, image segmentation, image matching, optimization, dynamic programming.

1. Introduction

Despite the dramatic variety of application problems of image analysis, it is possible to set off some single subclasses of problems which allow for treating them in the unified terms of respective standard mathematical optimization problems for which there exist effective methods of solving. One of such classes of image processing problems is considered in this paper.

We will consider an image as a numerical function $y_t, t = (t_1, t_2)$, usually, brightness, which takes values from an appropriate set $y \in \mathcal{Y}$, and is defined on the discrete image pixel grid $t \in T$.

In this work, we restrict our concern to only those of image processing problems which can be represented as problems of transforming the original image $Y = (y_t, t \in T), y \in \mathcal{Y}$, into another function $X = (x_t, t \in T)$, that would be defined on the same argument set $t \in T$ and take values $x \in \mathcal{X}$ from a set specific for each particular problem.

As samples of such problems, at least, four of them should be mentioned, namely, the problems of smoothing, segmentation, image matching, and building the map of local texture orientation. These problems were considered in [1] from the statistical viewpoint as those of estimating the hidden Markov component of a two-component random field, in which the original image acted as the observable component. The assumed Markov property of the hidden field $X = (x_t, t \in T)$ was meant to express the *a priori* smoothness constraints of that or other kind, which usually follow from the nature of the respective practical problem.

However, the essence of both advantages and difficulties of such an approach lies, actually, not in its probabilistic aspect, but in the fact that the procedures of estimating the realization of the hidden random field $\hat{X}(Y)$ are those of solving a specific optimization problem $\hat{X}(Y) = \arg \min(\textit{or } \max) J(X|Y)$. The specificity of the problem is that the variables constituting the vector to be evaluated $X = (x_t, t \in T)$ are associated with nodes of a neighborhood graph defined on the image pixel set, and that, due to the principal Markov assumption, the objective function $J(X|Y)$ is separable in accordance with this graph. This fact itself does not contribute essentially to an improvement of the computational solvability of the problem [2], until the neighborhood graph is assumed to have no cycles, i.e. to be a tree. It is just the latter assumption which leads to highly effective resulting algorithms and is systematically exploited within the bounds of the statistical approach to image analysis.

In particular, in [3] the usual lattice-like neighborhood graph is approximated by a combination of "horizontal" trees lying immediately in the image plane, and in [4] the neighborhood tree has the form of a "vertical" multiscale pyramid.

In this work, we consider the "horizontal" version of the generalized image processing problem specified above as that of finding the minimum of a separable objective function $J(X|Y)$ defined on the pixel neighborhood graph, without reference to any probabilistic interpretation. Such a function does not lend itself to an easy numerical minimization in the general case, therefore, we decompose it into a combination of partial objective functions each defined on a tree-like pixel-neighborhood subgraph that rests on all the pixels as its nodes.

The global minimization of each of the tree-supported separable objective functions is provided by a generalized dynamic programming procedure. Besides, we consider here two particular versions of such a procedure, when the set $x_t \in \mathscr{X}$ the hidden variables take values from is finite $x_t \in \mathscr{X} = \{1, \ldots m\}$, and when it is the n-dimensional real space $x_t \in \mathscr{X} = \mathsf{R}^n$ with quadratic constituents of the separable objective functions. These versions are illustrated, respectively, by the problems of texture image segmentation and image matching.

2. The Generalized Image Processing Problem as that of Minimization of a Separable Objective Function Defined on the Pixel-Neighborhood Graph

As a rule, the essence of the particular image processing problem suggests a way to preliminarily judge, in a fuzzy form, about the value of the hidden field $x_t \in \mathscr{X}$ at a single pixel t (node of the pixel-neighborhood graph) from some vicinity Y_t of this point in the original image Y. It is convenient to express such a judgment in the form of a function $\psi_t^o(x_t|Y)$ which is defined over the set $\mathscr{X}$ and takes the greater value the more evident is the contradiction between the hypothesis that $x_t \in \mathscr{X}$ is just the "correct" local value we are seeking and the current image fragment Y_t. Let such a function $\psi_t^o(x_t | Y)$ be called image-dependent node function, with the upper index "o" meant as *observation-based*.

However, practically inevitable interferences of various kinds usually prevent from taking the totality of locally optimal values $\hat{x}_t(Y_t) = \arg \min \psi_t^o(x | Y_t)$, $x \in \mathscr{X}$, as the final decision on the sought-for hidden field $\hat{X}(Y) = [\hat{x}_t(Y_t), t \in T]$.

On the other hand, the same essence of the particular application problem suggests, as a rule, also a fuzzy *a priori* knowledge on the expected form of the hidden field. This knowledge often can be expressed in local terms as a function $\gamma_{t',t''}^m(x_{t'}, x_{t''})$ which is defined over $\mathscr{X} \times \mathscr{X}$ at each pair of neighboring pixels and takes the greater value the greater is the discrepancy between the respective hidden values. Let function $\gamma_{t',t''}^m(x_{t'}, x_{t''})$ be called *model-based* edge function. The pixel neighborhood depends on the particular problem, and is expressed, in the general case, by an undirected graph $G \in T \times T$.

Given an image Y, the decision on the hidden field $X = (x_t, t \in T)$ should be sought for as a compromise between the mutually contradictory requirements to minimize both node and edge functions. It is natural to take the sum

$$J(X|Y) = \sum_{t \in T} \psi_t^o(x_t | Y_t) + \sum_{(t',t'') \in G} \gamma_{t',t''}^m(x_{t'}, x_{t''}) \qquad (1)$$

as a combined criterion of decision making. Such a criterion is called here G-supported separable objective function. The number of its variables is equal to the number $|T|$ of nodes in G, i.e. the number of pixels.

Unfortunately, the intent to create a non-iterative procedure of finding the minimum point $\hat{X}$ of a separable function (1) on a supporting graph of general kind comes up against a fundamental barrier. The simplest form of the pixel neighborhood relation usually considered to be appropriate for images is lattice (Fig. 1a). Such a graph does not possess any specific properties from the viewpoint of minimizing a separable function in comparison with the general case.

But in the particular case, when the graph has no cycles, i.e. is a tree, there exists a highly effective global optimization procedure, based on a recurrent decomposition of the initial problem of minimizing function (1) of $|T|$ variables into a succession of $|T|$ partial problems, each of which consists in minimizing a function of only one variable $x \in \mathscr{X}$.

Below, in Section 5, we replace the neighborhood lattice by a combination of trees, each of which has no additional branching nodes except those of the stem, and differs from the others only by the position of this stem (Fig. 1b).

In a still more particular case, when tree G has no branching nodes, i.e. is a chain (Fig. 1c), such a procedure is nothing else than a version of the famous dynamic programming procedure.

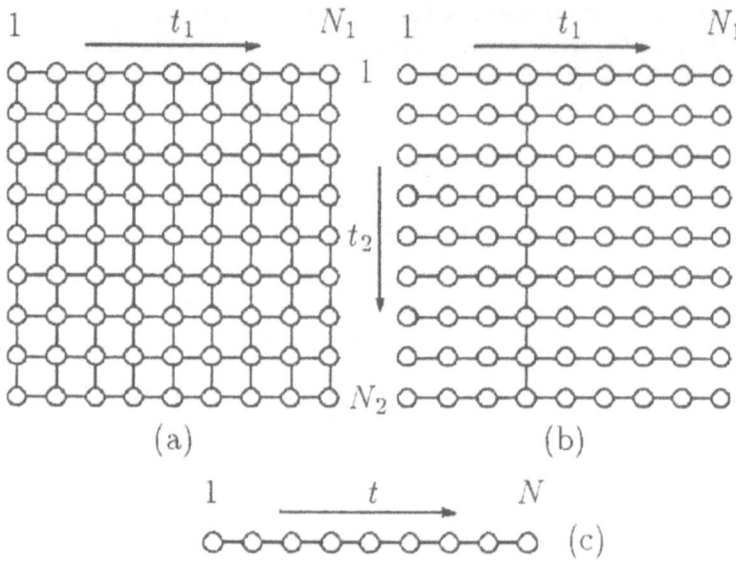

Figure 1. Neighborhood graphs on the pixel grid: a rectangular lattice (**a**), the simplest tree (**b**), and a chain (**c**)

3. The Basic Optimization Procedure for a Separable Function Supported by a Tree

In this and the next Section, we do not associate the variables of the objective function, i.e. the nodes of the supporting tree, with variables in any certain application problem. Let G be a tree on the set of nodes $T = \{t\}$ (Fig. 2), and

$$J(X) = \sum_{t \in T} \psi_t^o(x_t) + \sum_{(t', t'') \in G} \gamma_{t', t''}^m(x_{t'}, x_{t''}), \quad x_t \in \mathcal{X}, \tag{2}$$

be a G-supported separable objective function formed by, respectively, node $\psi_t^o(x_t)$ and edge functions $\gamma_{t', t''}^m(x_{t'}, x_{t''})$.

It is out of significance which of the nodes t^* is assigned the root of the tree, but once it is done, the node set gets partitioned into a number of subsets each consisting of the nodes of a certain level (Fig. 2):

$$T = \bigcup_{j=1}^{M} T^j, \quad T^j \cap T^k = \varnothing, \quad j \neq k.$$

We will denote the set of the nodes of the lowest level by T^1, and the set containing the only node of the highest level, namely, the root, by T^M.

Let, further, the tree $G_t \subset G$ formed by a node t and its descendants be called descendant tree of this node. It is clear that $G_t = \varnothing$ for the leaves, in particular, for all $t \in T^1$, and that $G_t = G$ for the root $t \in T^M$. The set of all the nodes in

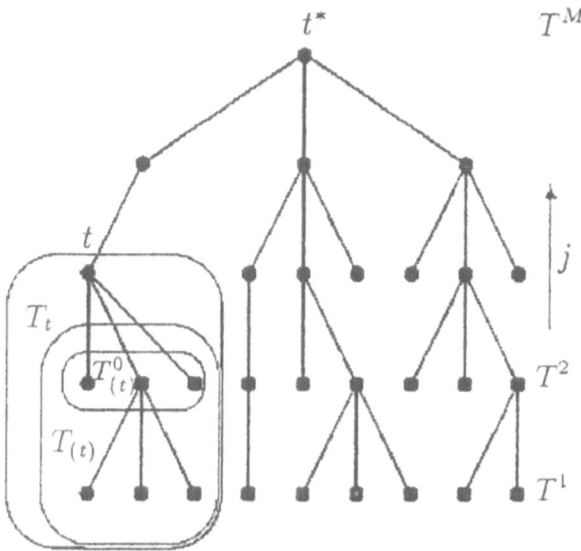

Figure 2. Structure of an arbitrary tree

the descendant tree of a node t will be denoted by T_t, symbol $T_{(t)}$ will mean the same set without this node itself, i.e. the set of its descendants, and $T^0_{(t)} \subseteq T_{(t)}$ will denote the subset of its immediate (next) descendants. So, $T_t = \{t\}$ and $T^0_{(t)} = T_{(t)} = \varnothing$ for the leaves, and $T_{t^*} = T$ for the root.

Analogously, symbols $X_t = (x_s, s \in T_t)$ and $X_{(t)} = (x_s, s \in T_{(t)})$ will mean the respective partial vectors of the node variables.

Finally, let us associate with each node t a partial objective function

$$J_t(X_t) = J_t(x_t, X_{(t)}) = \sum_{s \in T_t} \psi^o_s(x_s) + \sum_{(s',s'') \in G_t} \gamma^m_{s',s''}(x_{s'}, x_{s''})$$

which has absolutely the same structure as the full objective function (2) but is defined on the respective partial descendant tree. It is easy to see that

$$J_t(X_t) = J_t(x_t, X_{(t)}) = \psi^o_t(x_t) + \sum_{s \in T^0_{(t)}} \{\gamma^m_{t,s}(x_t, x_s) + J_s(x_s, X_{(s)})\}.$$

In particular, for the nodes of the lowest level $t \in T^1$, as well as for all the leaves, the partial objective function gets trivial $J_t(X_t) = J_t(x_t) = \psi^o_t(x_t)$, and for the root the partial function is nothing else than the full objective function $J_{t^*}(X_{t^*}) = J(X)$ (2).

The principal idea of the optimization procedure for separable objective functions supported by trees is based, like the classical dynamic programming procedure for chains, on the notion of the Bellman function

$$\tilde{J}_t(x_t) = \min_{X_{(t)}} J(x_t, X_{(t)}) = \psi^o_t(x_t) + \min_{X_s, s \in T^0_{(t)}} \sum_{s \in T^0_{(t)}} \{\gamma^m_{t,s}(x_t, x_s) + J_s(X_s)\}, \quad x_t \in \mathcal{X}.$$

The fundamental property of the Bellman function

$$\tilde{J}_t(x_t) = \psi_t^o(x_t) + \sum_{s \in T_{(t)}^0} \min_{x_s \in \mathcal{X}}\{\gamma_{t,s}^m(x_t, x_s) + \tilde{J}_s(x_s)\} \tag{3}$$

will be called the upward recurrent relation. The inverted form of this relation

$$\tilde{x}_s(x_t) = \arg \min_{x_s \in \mathcal{X}}\{\gamma_{t,s}^m(x_t, x_s) + \tilde{J}_s(x_s)\}, \quad s \in T_{(t)}^0, \tag{4}$$

will be referred to as the downward recurrent relation.

The optimization procedure is based on the assumption that there exists a sufficiently effective way of solving the partial optimization problems occurring in (3), and storing, in a compact form, the Bellman functions or downward relations.

The procedure runs twice through all the nodes of the graph in order of their membership in the hierarchy levels T^j, first upwards $j = 1, \ldots, M$, and then downwards $j = M, \ldots, 2$.

It starts with the nodes of the lowest (first) hierarchy level $t \in T^1$, $j = 1$. As $T_{(t)}^0 = \emptyset$ for these nodes,

$$\tilde{J}_t(x_t) = \psi_t^o(x_t), \quad x_t \in \mathcal{X}, \quad t \in T^1.$$

Thereupon, the upward recurrent relation (3) is applied sequentially to the nodes of the higher levels $j = 2, \ldots, M$. The Bellman functions $\tilde{J}_t(x_t)$ or downward relations $\tilde{x}_s(x_t)$, $x_t \in \mathcal{X}$, are to be stored for all the nodes of the tree except those of the first level.

The Bellman function at the root $\tilde{J}_{t^*}(x_{t^*})$, $x_{t^*} \in \mathcal{X}$, that is obtained at the last step of the upward pass, immediately gives the optimal value of the root variable:

$$\hat{x}_{t^*} = \arg \min_{x_{t^*} \in \mathcal{X}} \tilde{J}_{t^*}(x_{t^*}).$$

On the downward pass, as the procedure descends the hierarchy levels $j = M, \ldots, 2$, the already found optimal values of the current level determine optimal values of the variables at the immediately underlying level in accordance with the downward recurrent rules (4), stored as the result of the upward pass or determined by the stored Bellman functions (3)

$$\hat{x}_s = \tilde{x}_s(\hat{x}_t), \quad s \in T_{(t)}^0.$$

The degenerate particular case of a tree is tree without branching nodes, i.e. a chain. In this case, the above described optimization procedure becomes the classical dynamic programming procedure.

We introduce here an additional denotation $\hat{J}_t(x_t)$ for the so-called marginal node functions each of which shows how the full objective function depends on the value of a single variable at the respective node t, if other node variables take the conditionally optimal values in accordance with x_t:

$$\hat{J}_t(x_t) \min_{x_s, s \in T, s \neq t} J(X).$$

At the root, the marginal node function is found at the last step of the upward pass $\hat{J}_{t^*}(x_{t^*}) = \tilde{J}_{t^*}(x_{t^*})$. It can be shown that the marginal node functions at other nodes, each of which inevitably is an immediate descendant $s \in T_{(t)}^0$ of a node t of the higher level, satisfy the downward recurrent relation

$$\hat{J}_s(x_s) = \min_{x_t \in \mathcal{X}}\{\hat{J}_t(x_t) + [\tilde{J}_s(x_s) + \gamma_{t,s}^m(x_t, x_s)] - \min_{x_s' \in \mathcal{X}}[\tilde{J}_s(x_s') + \gamma_{t,s}^m(x_t, x_s')]\}. \quad (5)$$

4. Numerical Realization of the Basic Procedure for Discretely Defined and Quadratic Node and Edge Functions

A numerical realization of the basic procedure for a separable function supported by a tree, just like the classical dynamic programming procedure, is possible only in case there exists a finitely parametrized function family $\tilde{J}(x; \mathbf{a})$ concordant with node functions $\psi_t^o(x)$ and edge functions $\gamma_{t',t''}^m(x', x'')$ in the sense that the Bellman functions (3) belong to this family at each step. In this case, the upward pass of the procedure consists in a recurrent re-evaluation of parameters $\tilde{\mathbf{a}}_t$ that completely represent the Bellman functions. We consider here two kinds of such a parametrization which present, in some sense, two extremes of the range of possible ways to define node and edge functions.

The evident case of a trivial parametrization is the case of a discrete range of node variables $\mathcal{X} = \{1, \ldots, m\}$. In this case, Bellman functions (3) are m-dimensional vectors, downward relations (4) are vectors, too, and the elementary enumeration gives an easy solution of the partial optimization problems.

Another extreme case is the assumption that the range of node variables is n-dimensional real space $\mathbf{x} \in \mathcal{X} = \mathsf{R}^n$, and both node and edge functions are quadratic with positive definite Hessian matrices:

$$\begin{aligned} \psi_t^o(\mathbf{x}) &= (\mathbf{x} - \mathbf{x}_t^0)^T \mathbf{Q}_t^0 (\mathbf{x} - \mathbf{x}_t^0), \\ \gamma_{t',t''}^m(\mathbf{x}', \mathbf{x}'') &= (\mathbf{x}' - \mathbf{x}'')^T \mathbf{U}_{t',t''}(\mathbf{x}' - \mathbf{x}''). \end{aligned} \quad (6)$$

It can be shown that, in accordance with (3), the Bellman functions will be likewise quadratic

$$\tilde{J}_t(\mathbf{x}_t) = \text{const} + (\mathbf{x} - \tilde{\mathbf{x}}_t)^T \tilde{\mathbf{Q}}_t (\mathbf{x} - \tilde{\mathbf{x}}_t), \quad \mathbf{x} \in \mathsf{R}^n,$$

and, so, completely represented by their minimum points $\tilde{\mathbf{x}}_t \in \mathsf{R}^n$ and positive definite Hessian matrices $\tilde{\mathbf{Q}}_t(n \times n)$ to be found on the upward pass:

$$\tilde{\mathbf{Q}}_t = \mathbf{Q}_t^0 + \sum_{s \in T_{(t)}^0} [\tilde{\mathbf{Q}}_s^{-1} + \mathbf{U}_{t,s}^{-1}]^{-1}, \quad \tilde{\mathbf{x}}_t = \tilde{\mathbf{Q}}_t^{-1}\left\{\mathbf{Q}_t^0 \mathbf{x}_t^0 + \sum_{s \in T_{(t)}^0} [\tilde{\mathbf{Q}}_s^{-1} + \mathbf{U}_{t,s}^{-1}]^{-1} \tilde{\mathbf{x}}_s\right\}. \quad (7)$$

Analogously, it can be derived from (4), (6), and (7), that the downward recurrent relations will be linear

$$\tilde{\mathbf{x}}_s(\mathbf{x}_t) = (\mathbf{I} - \tilde{\mathbf{H}}_s)\tilde{\mathbf{x}}_s + \tilde{\mathbf{H}}_s \mathbf{x}_t = \tilde{\mathbf{x}}_s + \tilde{\mathbf{H}}_s(\mathbf{x}_t - \tilde{\mathbf{x}}_s), \quad s \in T_{(t)}^0,$$

$$\tilde{\mathbf{H}}_s = (\tilde{\mathbf{Q}}_s + \mathbf{U}_{t,s})^{-1}\mathbf{U}_{t,s} = [\mathbf{U}_{t,s}^{-1}\tilde{\mathbf{Q}}_s + \mathbf{I}]^{-1}.$$

Finally, all the marginal functions will be quadratic functions

$$\hat{J}_t(\mathbf{x}_t) = \text{const} + (\mathbf{x}_t - \hat{\mathbf{x}}_t)^T \hat{\mathbf{Q}}_t(\mathbf{x}_t - \hat{\mathbf{x}}_t)$$

defined by the final optimal values of the node variables and marginal Hessian matrices, both to be found on the downward pass in accordance with (5):

$$\hat{\mathbf{x}}_s = \tilde{\mathbf{x}}_s(\hat{\mathbf{x}}_t), \quad \hat{\mathbf{Q}}_s = [\tilde{\mathbf{H}}_s \hat{\mathbf{Q}}_t^{-1}\tilde{\mathbf{H}}_s^T + (\tilde{\mathbf{Q}}_s + \mathbf{U}_{t,s})^{-1}]^{-1}, \quad s \in T_{(t)}^0.$$

5. Tree Approximation of the Image Pixel Lattice and the Generalized Image Processing Technique

It is clear that there is no way to replace the lattice-like neighborhood graph (Fig. 1a) by a tree-like one without loss in the ability to ensure smoothness of the secondary data field in all directions from each point of the image plane.

To avoid this obstacle, for finding the values of the hidden field at each vertical row of the picture, we use a separate pixel neighborhood tree which is defined, nevertheless, on the whole pixel set and has the same horizontal branches as the others (Fig. 1b). The resulting image processing procedure is aimed at finding the optimal values only for the hidden variables at the stem nodes in each tree, and boils down to a combination of two usual dynamic programming procedures, each dealing with single image rows, respectively, the horizontal and vertical ones. First, such a one-dimensional procedure is applied to the horizontal rows $t_1 = 1, \ldots, N_1$ independently for each $t_2 = 1, \ldots, N_2$. The resulting marginal node functions $\hat{J}_{t_1,t_2}(x)$ have to be stored in the memory. Then, the procedure is applied to the vertical rows $t_2 = 1, \ldots, N_2$ independently for each $t_1 = 1, \ldots, N_1$ with the only alteration: the respective marginal node functions $\hat{J}_t(x)$, obtained at the first step, are taken instead of the image-dependent node functions $\psi_t^o(x)$.

6. Discrete Optimization: Segmentation of Seismic Sections Considered as Texture Images

In the course of gas and oil reserves exploration, the so-called seismic sections are analyzed, that consist of synchronous records of reflected seismic signals registered by a large number of geophones (seismic sensors) placed along a straight line on the earth surface (Fig. 3). As the source of the initial seismic pulse, usually serves a series of explosions, responses to which are averaged in a special manner. Although the initial argument of the seismic signals is time, a relevant preprocessing allows to identify the vertical axis of the picture with depth under the respec-

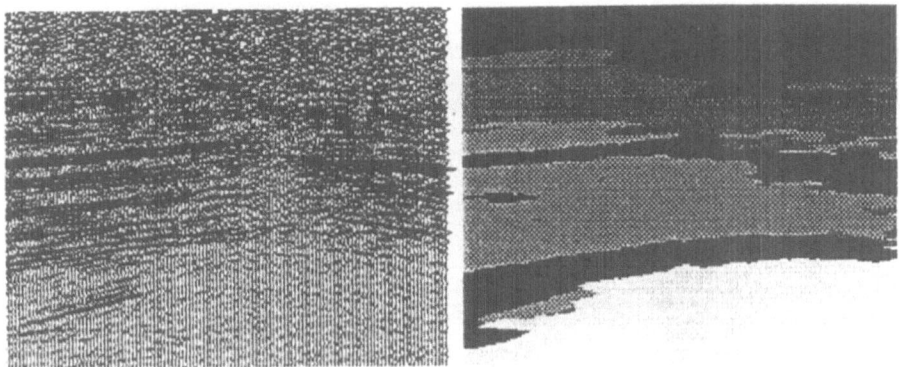

Figure 3. A vertical seismic section and the result of its segmentation into rock layers of six classes

tive sensor. Segmentation of a seismic section as a specific texture image is aimed at building a geological model of the layered rock mass.

In the problem of segmentation, the class allocation of the image pixels will be naturally expressed by a variable taking values from a finite set of class-membership indices $\mathscr{X} = \{1, \ldots, m\}$. So, the segmentation of the original image can be sought for in the form of a class allocation map $X = (x_t, t \in T)$.

Let us assume that the texture models are already chosen for all the classes of the sought-for homogeneous regions $x \in \mathscr{X} = \{1, \ldots, m\}$ and expressed each in the form of a linear autoregression model $y_t \cong \Sigma a_i^x y_{t+i}, i \in I^0, i \neq 0$, where the set of shifts I^0 determines the current image fragment Y_t. Then, the image-dependent node functions, discrete in this case, can be taken in the form

$$\psi_t^o(x \mid Y_t) = \left(y_t - \sum_{i \in I^0, i \neq 0} a_i^x y_{t+i} \right)^2, \quad x \in \mathscr{X} = \{1, \ldots, m\}.$$

As the simplest version of the model-based edge functions, meant to express the discrepancy between two neighboring values of the discrete hidden class-membership index, we used coincidence. functions $\gamma_{t',t''}(x_{t'}, x_{t'}) = 0$ if $x_{t'} = x_{t'}$, and $\gamma_{t',t''}(x_{t'}, x_{t'}) = \alpha_{t',t''}$ if $x_{t'} \neq x_{t''}$ with much greater value of $\alpha_{t',t''}$ at horizontal edges to emphasize the expected horizontally stretched form of the sought-for rock layers. The result of segmentation is shown in Fig. 3.

7. Quadratic Optimization: Matching of Two Images of Similar Structure

Let two images $Y^l = (y_t^l, t \in T)$, $Y^r = (y_t^r, t \in T)$ have the same range of definition T and form an entire image pair $Y = (Y^l, Y^r)$ to be analyzed as a whole, for instance, the left and the right image of a stereo pair, or photos of two human faces taken on the same scale (Fig. 4), and the like. Whatever objective might be pursued by the analysis of a pair of similar images, the first and algorithmically

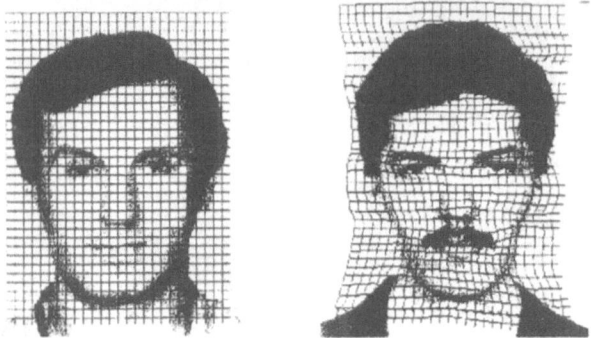

Figure 4. A pair of human portraits and disparity map of their identical points

most difficult stage is matching, i.e. establishing a correspondence in the two images between the point pairs which could be treated as identical. The result of such an operation can be represented in the form of the so-called disparity map which, for each point t of the one of the images (conventionally taken as reference image), determines vector $\mathbf{x}_t = (x_t^1, x_t^2) \in R^2$ indicating the spatial shift between t and its match $t' = t + \mathbf{x}_t$ in the second (search) image with eventual interpixel interpolation [5]. Thus, the result of matching of two images $Y = (Y^l, Y^r)$ should be represented as a secondary vector data field (disparity map) $\mathbf{X} = (\mathbf{x}_t, t \in T)$, $\mathbf{x} \in \mathcal{X} = R^2$, with the same range of definition as the reference image.

As universally adopted, a reference (left) picture element t, represented by some vector of its local features $\mathbf{z}_t^l$, is to be compared with the set of analogously represented search (right) picture elements $\mathbf{z}_{t+\mathbf{x}}^r$ accessible from t by admissible shifts $\mathbf{x}$ forming the elementary fragment Y_t of the image pair. The quadratic approximation of the squared Euclidean norm of the vector feature difference within the admissible shift set

$$\psi_t^o(\mathbf{x}) = c + (\mathbf{x} - \mathbf{x}_t^o)^T \mathbf{Q}_t^o (\mathbf{x} - \mathbf{x}_t^o) \cong \|\mathbf{z}_t^l - \mathbf{z}_{t+\mathbf{x}}^r\|^2$$

will be an appropriate image-dependent node function in which the eigenvectors and eigenvalues of $\mathbf{Q}_t^o$ carry an information on the plausibility of deflections of the local decision $\mathbf{x}_t$ in different directions from the most likely value $\mathbf{x}_t^o$.

8. Conclusions

Many image processing problems may be represented as those of transformation of the original image into a secondary data field on the same pixel grid. The essence of the problem consists in coordinating the mutually contradictory image-dependent information and model-based smoothness constraints both defined on the pixel neighborhood graph. At the same time, it is just this aspect of the image processing techniques which implies the major part of the amount of processing operations. We have shown that such a generalized problem can be solved by a high-speed procedure of minimizing a separable function defined on a succession of pixel neighborhood trees. The specificity of each particular image processing problem falls only on the way of extracting single units of the local information on the hidden field immediately from the original image, i.e. on the forming of an appropriate image-dependent node function.

References

[1] Mottl, V. V., Muchnik, I. B., Blinov, A. B., Kopylov, A. V.: Hidden tree-like quasi-Markov model and generalized technique for a class of image processing problems. 13th International Conference on Pattern Recognition. Vienna, Austria, August 25–29, 1996. Track B, pp. 715–719.

[2] Geman, S., Geman, D.: Stochastic relaxation, Gibbs distributions, and the Bayesian restoration of images. IEEE Trans. PAMI 6, 721–741 (1984).

[3] Wu, C.-H., Doershuk, P. C.: Tree approximation to Markov random fields. IEEE Trans. Patt. Anal. Mach. Intel. 17, 391–402 (1995).

[4] Luettgen, M. R., Karl, W. C., Willsky, A. S., Tenney, R. R.: Multiscale representation of Markov random fields. IEEE Trans. Sign. Proc. 41, 3377–3396 (1993).

[5] Mottl, V. V., Kopylov, A. V., Blinov, A. B., Zheltov, S. Yu.: Quasi-statistical approach to the problem of stereo image matching. Proc. SPIE 2363, 50–61 (1994).

V. V. Mottl
A. B. Blinov
A. V. Kopylov
A. A. Kostin
Tula State University
Lenin Ave. 92
300600 Tula
Russia
e-mail: mottl@atm.tsu.tula.ru

SpringerComputerScience

W. Kropatsch, R. Klette, F. Solina in cooperation with R. Albrecht (eds.)

Theoretical Foundations of Computer Vision

1996. 87 figures. VII, 256 pages. ISBN 3-211-82730-7

Soft cover DM 165,–, öS 1155,–, US $ 119.00

Reduced price for subscribers to "Computing":

Soft cover DM 148,50, öS 1039,50

Computing / Supplement 11

Computer Vision is a rapidly growing field of research investigating computational and algorithmic issues associated with image acquisition, processing, and understanding. It serves tasks like manipulation, recognition, mobility, and communication in diverse application areas such as manufacturing, robotics, medicine, security and virtual reality. This volume contains a selection of papers devoted to theoretical foundations of computer vision covering a broad range of fields, e.g. motion analysis, discrete geometry, computational aspects of vision processes, models, morphology, invariance, image compression, 3D reconstruction of shape. Several issues have been identified to be of essential interest to the community: non-linear operators; the transition between continuous to discrete representations; a new calculus of non-orthogonal partially dependent systems.

Springer Wien New York

Sachsenplatz 4-6, P.O.Box 89, A-1201 Wien, Fax +43-1-330 24 26
e-mail: order@springer.at, Internet: http://www.springer.at
New York, NY 10010, 175 Fifth Avenue • D-14197 Berlin, Heidelberger Platz 3
Tokyo 113, 3-13, Hongo 3-chome, Bunkyo-ku

SpringerJournals

Computing

Archives for Informatics and Numerical Computation

Editorial Board:

R. Albrecht, Innsbruck; H. Brunner, St. John's;
R. E. Burkard, Graz; W. Hackbusch, Kiel;
G. R. Johnson, Fort Collins; W. Knödel, Stuttgart;
W. G. Kropatsch, Wien; H. J. Stetter, Wien;
and an international Advisory Board

Computing publishes original papers and short communications from all fields of scientific computing in English (preferably) or German. Contributions may be of theoretical or applied nature, the essential criterion is computational relevance. Subject areas include discrete mathematics, symbolic computation, parallel computation, computer arithmetic, architectural concepts for computers and networks, operating systems, programming languages, software engineering, performance and complexity evaluation, data bases, image processing, computer graphics, pattern recognition, artificial intelligence, optimization, numerical analysis, and numerical statistics.

Subscription Information:

1998. Vols. 60+61 (4 issues each):

DM 1.168,–, öS 8.176,–, plus carriage charges

ISSN 0010-485X, Title No. 607

For customers in EU countries without VAT identification number

10 % VAT will be added to the subscription price

SpringerWienNewYork

Sachsenplatz 4-6, P.O.Box 89, A-1201 Wien, Fax +43-1-330 24 26
e-mail: order@springer.at, Internet: http://www.springer.at
New York, NY 10010, 175 Fifth Avenue • D-14197 Berlin, Heidelberger Platz 3
Tokyo 113, 3-13, Hongo 3-chome, Bunkyo-ku

Springer-Verlag
and the Environment

WE AT SPRINGER-VERLAG FIRMLY BELIEVE THAT AN international science publisher has a special obligation to the environment, and our corporate policies consistently reflect this conviction.

WE ALSO EXPECT OUR BUSINESS PARTNERS – PRINTERS, paper mills, packaging manufacturers, etc. – to commit themselves to using environmentally friendly materials and production processes.

THE PAPER IN THIS BOOK IS MADE FROM NO-CHLORINE pulp and is acid free, in conformance with international standards for paper permanency.